Okamoto

Microcellular Processing

Kelvin T. Okamoto

Microcellular Processing

HANSER
Hanser Publishers, Munich
Hanser Gardner Publications, Inc., Cincinnati

The Author:
Dr. Kelvin T. Okamoto, 717 Atlantic Avenue #3 A, Boston, MA 02111

Distributed in the USA and in Canada by
Hanser Gardner Publications, Inc.
6915 Valley Avenue, Cincinnati, Ohio 45244-3029, USA
Fax: (513) 527-8801
Phone: (513) 527-8977 or 1-800-950-8977
Internet: http://www.hansergardner.com

Distributed in all other countries by
Carl Hanser Verlag
Postfach 86 04 20, 81631 München, Germany
Fax: +49 (89) 98 48 09
Internet: http://www.hanser.de

Library of Congress Cataloging-in-Publication Data

Okamoto, Kelvin T.
Microcellular processing / Kelvin T. Okamoto.-- 1st ed.
p. cm.
ISBN 1-56990-344-1 (Hardcover)
1. Plastic foams. I. Title.
TP1183.F6O38 2003
668.4'93--dc21
2003007495

Bibliografische Information Der Deutschen Bibliothek
Die Deutsche Bibliothek verzeichnet diese Publikation in der Deutschen Nationalbibliografie;
detaillierte bibliographische Daten sind im Internet über <http://dnb.ddb.de> abrufbar.

ISBN 3-466-22344-4

Production Management: Oswald Immel
Coverdesign: MCP • Susanne Kraus GbR, Holzkirchen, Germany
Typeset, printed and bound by Kösel, Kempten, Germany

Foreword

The birth of microcellular plastics owes its origin to the MIT-Industry Polymer Processing Program, which was originally funded in 1973 by the National Science Foundation (NSF) as an experiment in promoting cooperative research between universities and industry in the United States. Based on the successful outcome of this experiment, the program has been expanded by NSF to include more than 100 centers of its kind throughout the United States. The Wydler-Stevenson Innovation Act was partly based on this model for cooperative research.

The MIT-Industry Polymer Processing Program has undertaken a large number of research projects since its inception. The graduates of the Program are now leaders in polymer processing and related fields, both at universities and in industry. One of the research projects undertaken by the MIT Program was the creation of microcellular plastics. The idea for microcellular plastics was created in only a few minutes during a luncheon at Eastman Kodak with Gordon Brown. The idea came about in response to the suggestion of Mr. Brown that MIT develop a technology for Eastman Kodak, which can save materials without jeopardizing material properties and without changing the part geometry. Even though the initial idea of the morphology of the microcellular plastics was designed in a few minutes on a napkin at the luncheon, the development of the manufacturing process took several years.

The research on microcellular plastics took place at MIT in two separate phases. During the first phase, from 1980 to 1984, MIT developed a batch process. During the second phase, which spanned from 1989 to 1995, continuous processes were developed. The first student to work on the project was Jane Martini, who came to MIT for her graduate degree after earning an undergraduate degree from Bryn Mawr College. As a student she was smart, energetic, and creative. The basic concept behind the microcellular plastics process was the use of thermodynamic instability to nucleate a plethora of cells. After the basic feasibility was demonstrated, a succession of outstanding students – Jae Ryun Yoon, Jonathan Colton, Vipin Kumar, and others – worked on various aspects of the batch process. The second phase – the development of continuous processes – commenced after I returned to MIT from a tour of duty with the U.S. Government in 1988. Chul Park, Dan Baldwin, Sung W. Cha, and other students contributed to the development of the process and the understanding of the basic process physics.

Trexel, Inc. took over the microcellular plastics technology as the exclusive licensee of MIT and developed extrusion and injection molding processes for microcellular plastics. They adopted MuCell as the registered tradename for the microcellular plastics and its processes. Trexel would not exist today without the vision and the leadership of Alex d'Arbeloff, Chairman of Trexel, Inc., who has also become Chairman of the MIT Corporation; Mr. D'Arbeloff was the founder and former CEO of Teradyne, Inc. He brought to Trexel, Inc. an able management team under the leadership of David Bernstein. Before the takeover of Trexel by a new management team, Matt Pallaver provided leadership in this difficult period during the initial commercialization phase of microcellular plastics.

The MuCell process has become an important industrial technology because of the able technical team assembled at Trexel, Inc., who are highly skilled and knowledgeable technical leaders who have developed the MuCell technology as a viable industrial process. Using the results of academic research they were able to make industrially viable products and processes, which have been licensed to a large number of industrial firms worldwide. They deserve much credit for their invaluable contributions. The results of their efforts clearly took many years of hard work.

I personally derive a great deal of satisfaction in the outcome of our microcellular plastics endeavor. Not only was the original goal of saving materials realized but also a lot more. The MuCell process shortens cycle time, reduces the clamping tonnage of injection molding machines, and produces "residual stress free" parts that do not warp. Furthermore, the process does not use any hydrocarbons, making it an environmentally friendly process. Most of all, it is the tangible result of university-industry cooperation that supports the idea that both universities and industry can benefit when they cooperate under a carefully structured framework.

Finally, my special thanks and congratulations go to Kelvin Okamoto and his colleagues for putting together this valuable book. It will be a major reference book on this important technology. I imagine we will find this book on the shelves of many engineers and students engaged in the advancement of the polymer processing field.

Nam P. Suh
Sudbury, MA, U.S.A.
February 2003

Preface

Microcellular processing of thermoplastic materials has taken some time to become commercially viable. However, microcellular processing is fast gaining acceptance and is now used to make products for automotive, electronic, and other demanding end-use applications. Trexel, Inc. (Woburn, MA) has been the leader in thermoplastic microcellular processing by developing and licensing microcellular process technology for extrusion, injection molding, and blow molding under the MuCell® trade name, which is a registered trademark of Trexel.

This book has been pulled together from information publicly available. Unfortunately, this means that information proprietary to Trexel or other companies is not contained here. At times, details of screw designs, die designs, and process conditions may be considered proprietary information. To obtain details not given here, please contact the appropriate company or one of a growing number of consultants knowledgeable in microcellular processing.

The largest thanks for helping this book come into being goes to the staff at Trexel, Inc. David Bernstein was generous enough to allow me to take this book on as a personal project and be separate from my employment at Trexel. I thank Levi Kishbaugh, Kent Blizard, Jingyi Xu, and Mark Berry at Trexel for reviewing the content for accuracy. I also thank Prof. Nam Suh of the Massachusetts Institute of Technology and of Trexel's Board of Directors for providing the Foreword for this book. Unless noted, all photographs in this book are courtesy of Trexel, Inc.

Additionally, I thank Peter Kennedy of Moldflow Corporation (Ithaca, NY) for providing the basic text and figures for Chapter 7 "Microcellular Molding: Moldflow Modeling." Peter was instrumental in ensuring the content remained accurate during my revisions and editing.

This book would never have made it to press without the insistence and persistence of Christine Strohm of this book's publisher, Hanser-Gardner, and the technical help of Oswald Immel. I thank them for keeping after me.

Last but definitely not least, I thank my wife Jenny for knowing that I wanted to write this book and accepting the role of book-widow for a few months.

Kelvin Okamoto

Contents

1 Introduction

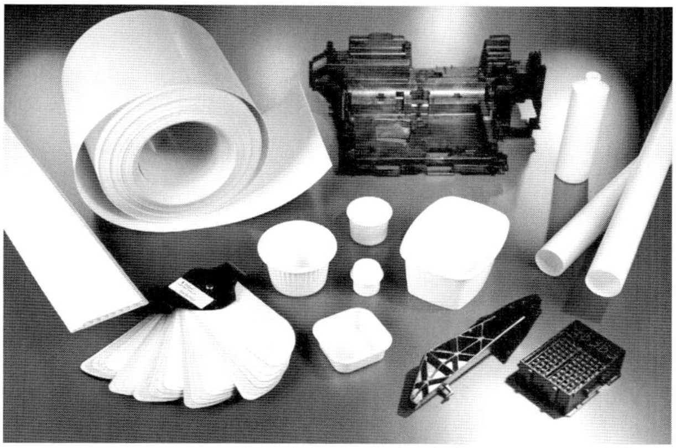

Figure 1.1 *Parts made using microcellular processing*

1.1 History

In the early 1980s Eastman Kodak was beginning to face heavy competition within its United States home market for highly profitable personal photographic film. As part of their effort to overcome this challenge, Eastman Kodak sought ways to reduce material costs and to improve manufacturing efficiency for its photographic film. Therefore Eastman Kodak approached Dr. Nam Suh, Mechanical Engineering Department Head of Massachusetts Institute of Technology (MIT) in Cambridge, Massachusetts, to help out and develop a new manufacturing technology that Eastman Kodak could implement.

The critical requirements given to Dr. Suh by Eastman Kodak included:

- Reduce plastics usage,
- Maintain physical properties, and
- Maintain or improve manufacturing efficiency.

Applying his axiomatic design technique Dr. Suh soon discovered that the first two criteria could be met by placing very small bubbles into the plastic. Furthermore, the tiny bubbles could be imparted into the plastics by rapidly decreasing the pressure or by rapidly increasing the temperature during plastic processing. The pursuit of this innovative idea using the basic theoretical guidelines given above, led to the invention of microcellular foam technology in the laboratories of Dr. Suh at MIT. Development of microcellular processing technology continued during the 1980s and early 1990s at MIT.

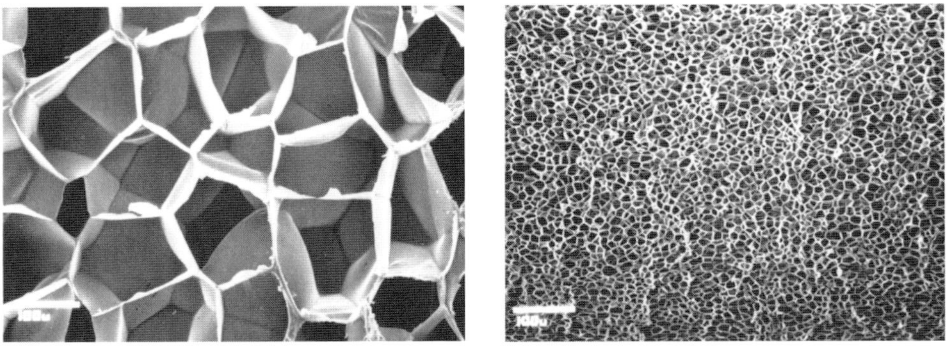

Figure 1. 2 *Left: Standard polystyrene foam*
Right: Microcellular polystyrene foam (batch process)

Commercial application of the breakthrough microcellular technology began in earnest during 1995 by Axiomatics Corp., which was later renamed Trexel, Inc., under exclusive patent rights from MIT. Axiomatics immediately encountered high commercial interest in the new technology and began the development using standard plastics processing equipment in cooperation with outside companies. Axiomatics also began to license the technology to other companies for use in their commercial processes.

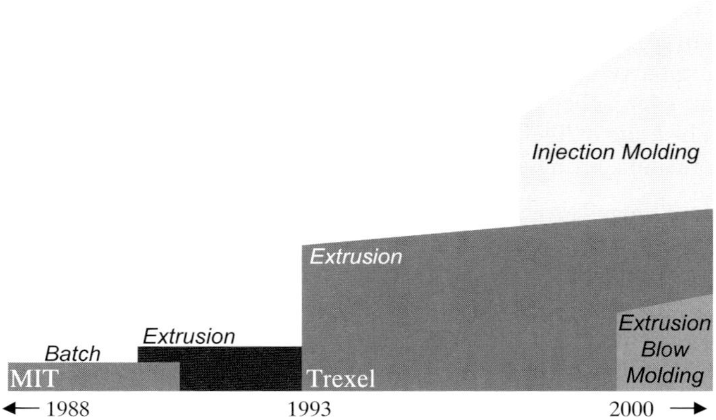

Figure 1.3 *Schematic timeline of microcellular processing development*

Today, Trexel is a company funded by private investment and venture capital with more than twenty employees. Trexel continues to develop the microcellular process for plastics extrusion, blow molding, and injection molding applications. Over one hundred companies now license the technology and many commercial products are manufactured using the microcellular process. Trexel calls the new technology the MuCell® microcellular process; MuCell and Trexel are registered trademarks of Trexel, Inc.

Through Trexel's licensing effort, numerous other organizations are also making further advances in thermoplastic microcellular processing. The list of organizations includes plastics original equipment manufacturers (OEMs), universities, and university/industry consortia. Numerous unlicensed organizations also are studying the microcellular process but may face difficulty licensing their own technology without first consulting with Trexel due to its existing intellectual property portfolio.

1.2 Patents Covering Microcellular Process Technology

The Massachusetts Institute of Technology and Dr. Suh patented their inventions in microcellular process technology throughout their development work. The first patent issued in the United States on September 25, 1984. Since then 7 more US patents and 4 international patents have issued. A list of MIT's microcellular patents and published applications are listed in Appendix 1; all of MIT's intellectual property regarding microcellular technology is exclusively licensed to Trexel.

As of January 2003, Trexel has continued to file patent applications worldwide to protect its intellectual property. Trexel itself holds 7 US patents and 5 international patents. A list of Trexel's patents and published applications are listed in Appendix 2.

Numerous other organizations have also filed patent applications regarding specific aspects of microcellular process technology. Some of the patent applications relate to a specific implementation of the process, e.g., in golf balls, and others relate to specific equipment or system designs. A partial list of patents by others is listed in Appendix 3.

2 General Description of the MuCell® Process

2.1 Background

Foam products made from thermoplastic polymers have been available commercially for many years. Although these products have been successful in a number of markets, their expansion within existing markets or into new markets often has been impeded by certain limiting characteristics of conventional foaming technologies.

The most severe limiting factors in the production and application of conventional thermoplastic foams are the large size of the cells that characterize these materials and the lack of uniformity of those cells. Large, non-uniform cells decrease mechanical properties including strength, toughness, and fatigue, introduce brittleness, and necessitate relatively thick part cross sections to ensure that the cells are contained within the material. Thin cross sections are often flawed by breaks and holes. Conventional foam processes also can be inconsistent due to the inherent difficulty controlling the levels and release of the blowing agent especially with the use of chemical blowing agents (CBAs). In addition, many foam processes use flammable blowing agents, which require regulatory approvals for their use and release, special handling techniques, and long product storage times.

MIT and Trexel have developed the microcellular thermoplastic foaming process for commercial injection molding, extrusion, and blow molding applications unaffected by the factors that limit the use of conventionally foamed thermoplastic polymers. The innovative process uses high cell nucleation rates within the foaming material to create foams with evenly distributed and uniformly sized microscopic cells. Microcellular foam has cell diameters averaging 100 microns or less, typically between 5 to 50 microns in diameter, depending on the material and application.

Foam materials produced by this process offer improved consistency and homogeneity of cell structures, which can result in products with superior properties and uniformity. While the tensile modulus declines as a function of the square of the density reduction in most foamed material, a more linear relationship often exists in microcellular foams because of the very small homogeneous cell structures. In products with skin/core cross sections such as those made by injection molding, the stiffness of parts can be even greater than predicted simply by the density reduction.

2.2 The Microcellular Foam Technology

Microcellular foam is produced when the cell nucleation rate is both extremely high (orders of magnitude greater than with conventional foaming processes) and much greater than the diffusion rate of the blowing agent into cells (i.e., cell growth). Under these conditions, an extremely large number of cells will be created before any cell growth occurs. Consequently, when the blowing agent diffusion begins to dominate in the foam creation process, all cell sites will begin to grow at the same time and at approximately the same rate, resulting in a material characterized by a large number of evenly distributed, uniformly sized, microscopic cells.

Atmospheric gases, such as carbon dioxide (CO_2) and nitrogen (N_2) – which are both less expensive than other common blowing agents and unregulated – can be used as physical blowing agents to create this phenomenon. Importantly, the very high nucleation rates needed to produce microcellular foam can be achieved without the use of nucleation agents, such as talc or calcium carbonate.

2.3 Microcellular Cells: The Theoretical Solution

Microcellular foams are defined by having uniform cell diameters of less than 100 microns. This cell structure can only be obtained through the simultaneous generation of a high number of nucleation sites. The requirement for very high nucleation rates and the resultant creation of a large number of nucleation sites necessitate a fundamental change in the way in which the cells are nucleated.

Briefly, conventional foaming technologies use physical blowing agents with nucleating agents or chemical blowing agents that induce heterogeneous nucleation in the material at a fixed – and relatively small – number of sites. The actual number of cell sites is directly related to the quantity of nucleating agents added. Heterogeneous nucleation produces a foam material characterized by large, non-uniform cells (see Fig. 2.1). The lack of cell size uniformity results from a relatively slow rate of nucleation. Typically, conventional foams have cells of 250 microns to 1 mm in diameter.

A high nucleation rate is driven by a large thermodynamic instability. This large thermodynamic instability is achieved by first dissolving a blowing agent into the molten polymer at high temperature and high pressure thus creating a single-phase solution and then lowering the pressure quickly to below the saturation pressure. To reach the desired high rate of nucleation, the saturation level of the dissolved blowing agent and/or the rate at which the instability is achieved must be high.

The conditions required for nucleation are best illustrated by plotting the solubility of a blowing agent as a function of pressure in a typical polymer system. In the three-dimensional schematic shown as Fig. 2.2, the solubility of CO_2 or N_2 in a molten polymer has been plotted as a function of pressure and temperature. As shown, the solubility of the blowing agent increases with increasing pressure (line AB) and decreases with increasing temperature (line AC). Within the plasticating section of injection molding equipment or within an extruder (point B), the pressure and temperature are high and the solubility of the blowing agent is high. At Point B, the blowing agent is dissolved in the molten polymer if the amount of blowing agent is less than the saturation amount.

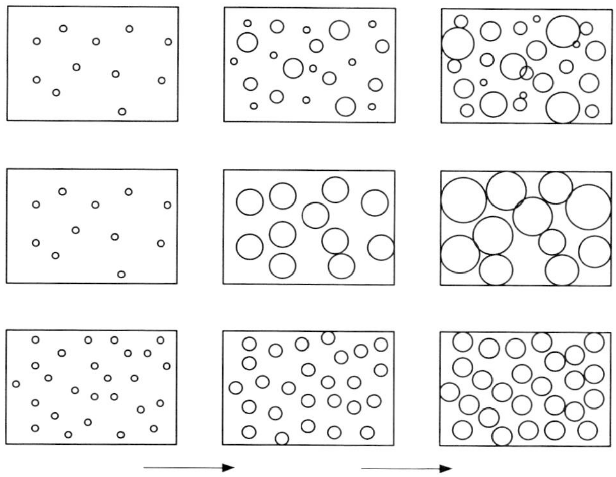

Figure 2.1 *Schematic showing the effect of homogeneous and heterogeneous nucleation and nucleation density on foam cell size and cell size distribution*

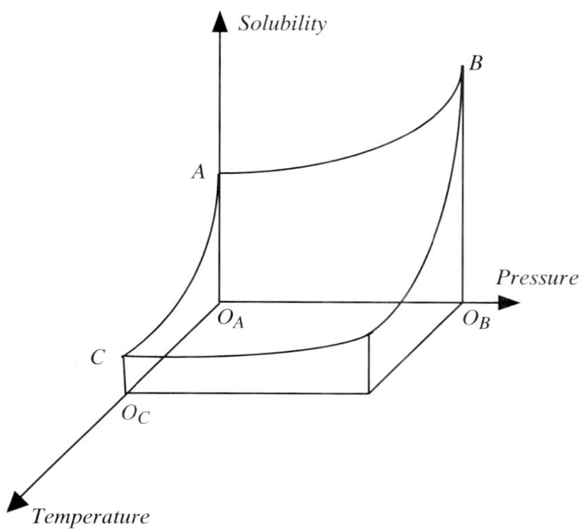

Figure 2.2 *Solubility of CO_2 or N_2 in a molten polymer*

In the die (point B to point A), the pressure drops quickly, and the blowing agent becomes supersaturated within the polymer; at this point, the blowing agent will begin to come out of the molten polymer as gas, foaming the polymer. If the drop in blowing agent solubility from A to B is sufficiently large and sufficiently fast, then conditions will exist for a high rate of nucleation of cells, and a large number of evenly distributed microscopic cells will form and grow uniformly.

2.4 Adaptation to Commercial Plastics Processing Equipment

Three conditions must reached during an injection molding, extrusion, or blow molding process for the required rapid homogeneous nucleation to take place:

- The appropriate amount of blowing agent (CO_2 or N_2) must be completely dissolved within the polymer – resulting in a single-phase solution – before the pressure is lowered.
- The pressure profile must be maintained throughout the system to keep the blowing agent in solution.
- A high rate of change of solubility must be established (accomplished by different means in extrusion and injection molding).

To promote the thermodynamic instability that drives the MuCell process, the blowing agent delivery system must be capable of transferring the requisite amount of blowing agent at a pressure higher than the pressure in the plasticating section of the injection molding equipment or in the metering section of the extruder. The creation of a single-phase solution by a diffusion controlled process depends on the diffusion coefficient, residence time, and temperature, as well as on the distance over which the blowing agent must diffuse. To facilitate the rapid creation of the single-phase solution, Trexel's MuCell process uses a physical blowing agent, e.g., CO_2, that is a supercritical fluid (SCF) to speed the process of dissolution. To achieve a uniform mix and dissolution of the required amounts of the blowing agent, the MuCell process employs a unique injection system and screw configuration designs.

Briefly, a supercritical fluid (SCF) is a material that is above both its critical pressure P_c and critical temperature T_c, as shown in Fig. 2.3. Generally, a hot gas can be made into an SCF by compression or a compressed liquid can be made into an SCF by heating. Carbon dioxide is an SCF above 71 bar and 31 °C while nitrogen is an SCF above 34 bar and –147 °C.

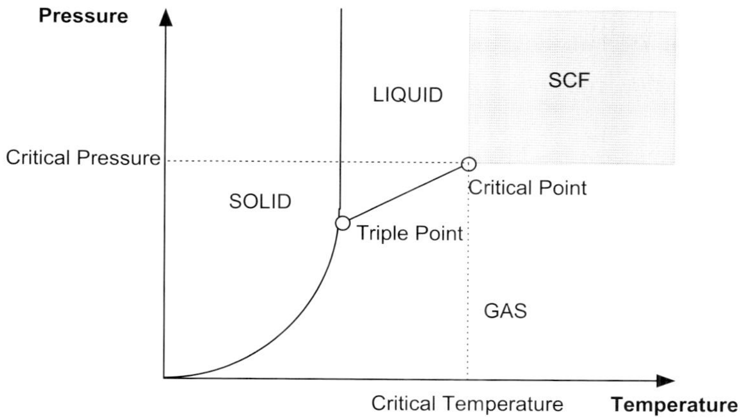

Figure 2.3 Material phase diagram

2.4.1 MuCell Molding Technology

In the plastication section of the injection molding process, a single-phase solution of blowing agent and polymer must be created. The single-phase solution is formed by applying knowledge about adding and mixing blowing agents in an extruder to the design of the plasticating portion of the injection molding system. An SCF metering system is required for this process as well as a specially designed SCF injection system for the extruder barrel. Special screw designs are also necessary for the complete development of a single-phase solution of blowing agent and polymer within the plasticating portion of the injection molding system.

Adapting injection molding equipment for the MuCell technology requires the integration of the MuCell process with the standard cyclic function of molding equipment. This integration, usually accomplished through software changes in the molding cycle, ensures that the correct amount of blowing agent for the polymer is delivered at the standard cycle of the equipment and that the single-phase, molten polymer/SCF solution is held at the conditions required to maintain the single-phase solution before it is delivered to the mold. To expose the single-phase solution to a pressure drop that is sufficiently rapid to cause nucleation, specific molding conditions must be maintained.

To summarize, adaptation of the microcellular process to injection molding requires the following changes or additions:

- An SCF metering system with sufficient capacity to deliver the blowing agent to the screw at the volume and pressure required for the MuCell process.
- An SCF injection system to properly inject the metered SCF into the plasticating section of the injection molding machine.
- A specially designed screw for creating a single-phase solution of blowing agent and polymer.

- Software and system modifications to create and maintain the single-phase solution throughout the injection molding cycle.
- Optimized molding conditions to create the microcellular foam within the molded part.

The microcellular process has been adapted and designed for both screw and accumulator injection molding machines and for standard reciprocating screw injection molding machines. The MuCell microcellular process can be retrofitted easily to existing injection molding machines. Furthermore, the MuCell microcellular process is available as an option on selected new injection molding machines from leading original equipment manufacturers. All MuCell-capable machines – new and retrofitted – are capable of conventional, non-foaming operation.

2.4.2 MuCell Extrusion Technology

To make an extrusion system compatible with the microcellular process, an SCF delivery system, SCF injectors, and modifications to the extruder screw and barrel are required. A die with the appropriate flow configuration also is necessary for implementation of the microcellular process, both to maintain the pressure required to keep the blowing agent in solution in the extruder and to create the high rate of change of solubility required for high nucleation rates.

To determine the proper die configuration, data describing the relationship between polymer viscosity, the melt temperature, and the blowing agent are used to calculate the pressure drop and pressure drop rate through the die flow path. Pressure develops as a viscoelastic fluid is pushed through a restrictive flow path at a known mass flow rate – i.e., in the MuCell process, pressure develops as the single-phase solution is pushed through the die. Using the pressure and pressure-drop rate requirements for high nucleation rates, the specific die flow geometry needed to produce microcellular foamed products can be designed. This process is covered by US Patent 6,051,174 by Park, et al. and other patents and patents pending.

To summarize, adaptation of the microcellular process to a conventional extrusion system requires the following changes or additions:

- An SCF metering system with sufficient capacity to deliver the blowing agent at the volume and pressure required.
- An SCF injector and screw designed to create a single-phase polymer solution.
- An extrusion die with flow paths to build pressure in the system and to provide the appropriate pressure drop rate

The MuCell technology can be retrofitted easily to installed extrusion equipment. All MuCell capable extruders – new and retrofitted – are also capable of conventional non-foaming or conventional foaming operation.

2.4.3 MuCell Blow Molding Technology

The modifications required to adapt blow molding equipment to the microcellular process – which are similar to the modifications required for the extrusion process – require modifications to the extruder to deliver SCF to the polymer melt and modifications to the blow molding die. For intermittent blow molding operations using a reciprocating screw, the modifications are similar to the modifications required for injection molding equipment.

2.5 Results and Advantages of the MuCell Process

2.5.1 MuCell Molding Technology

The MuCell process permits product designers to specify lightweight, molded foam materials for products, especially products thinner than 3.0 mm that cannot be manufactured with conventionally foamed materials and/or are impractical or cost prohibitive to manufacture with other lightweighting techniques such as gas-assist. With this breakthrough technology, three-dimensional objects with skin/core cross sections can be foamed with excellent dimensional tolerances. In many cases, the improved dimensional stability and rapid cooling of the part allows for cycle times to be reduced by as much as 10 to 50%. The MuCell process, which offers controlled weight reductions of up to 30% is suitable for thin walls down to 0.25 mm in thickness.

The use of SCFs as blowing agents lowers the viscosity of the melt up to 50%, which results in a substantial decrease in processing temperatures; the temperature reduction is as high as 80 °C. Other MuCell processing improvements include reductions in injection pressure up to 50%, improved mold filling, reductions of clamp tonnage up to 60%, elimination of sink marks, and reductions in post-mold warpage. The MuCell process can result in faster cycle times due to the reduction of melt temperature, improved mold filling, and the reduction in or elimination of hold time.

The low pressures associated with the MuCell process permit molders to increase the number of cavities without increasing clamp tonnage, which can significantly improve production efficiency. Likewise, larger parts can be molded without increasing clamp tonnage, which also can improve production efficiency. In addition, the low injection and clamp pressure requirements of the MuCell process permit molders to switch from steel molds to aluminum or other lower-cost materials and to more cost-effective methods of manufacturing.

2.5.2 MuCell Extrusion Technology

In extrusion applications, the MuCell process can be used to produce a variety of sheets, profiles, and coatings:

- Sheets made from microcellular foams are stable against additional processing, such as thermoforming or compression molding. The microscopic cells in MuCell material resist collapse and exhibit little or no elongation in foaming. Because of the microscopic cell size, sheets can be produced exhibiting significant weight savings over unfoamed products and still maintaining the appearance of solid materials.
- Profiles can be produced that exhibit weight reductions over unfoamed products, resulting in significant weight and cost savings. Because of the stability and uniformity of the microscopic cells, profiles can be subjected to additional forming techniques, such as blow molding tube profiles into hollow shapes. Due to the small cell size in microcellular foam, MuCell profiles have an appearance similar to solid materials.
- Coatings can be produced with extremely small dimensions – less than 6 mils.

MuCell extruded materials reflect many advantages for extruders:

- All MuCell foams are produced with a low-cost blowing agent, such as CO_2 or N_2.
- No additives (e.g., nucleation agents) are required to produce MuCell microcellular foams.
- No reactive components (e.g., CBAs) or viscoelasticity modifiers must be added to the materials.
- The control of the foaming process is built into the die and extruder system and does not require some of the process control features normally found in foam extrusion systems.

2.5.3 MuCell Blow Molding Technology

In blow molding applications, the primary advantage of the MuCell foam operation is the reduction of material consumption. The MuCell process has been used to produce HDPE parisons (the tubular precursor to the blown bottle) with density reductions up to 35%. Bottles blown from these parisons retain this density reduction, despite the high blow pressures required to form the parison into the bottle shape.

2.6 Summary

Microcellular foams can be made using injection molding or extrusion equipment at commercial production rates with low-cost modifications. The key to producing microcellular foams is the use of rapid nucleation techniques and the control of the pressure and pressure-drop rate in injection nozzles and extrusion dies.

The MuCell microcellular foam process provides many new application possibilities to molders and extruders. In particular, materials that have been difficult to foam and/or require CBAs with conventional foaming processes can be produced with the MuCell process because of the precise process control inherent in this technology. In addition, the microscopic cell sizes inherent in MuCell material permit the successful extrusion and molding of very thin parts.

3 SCF Blowing Agent Delivery Systems

3.1 Background

Microcellular processing utilizes physical blowing agents injected into a molten polymer stream while in a supercritical state. This chapter covers the specifications for the physical blowing agents as well as the equipment used to deliver the supercritical fluid (SCF) to the extruder or plasticator barrel.

3.2 Physical Blowing Agent Specification

Most stable chemicals that are liquids or gases at standard temperature and pressure may be used in microcellular processing. Physical blowing agents that have been reported to be used in microcellular processing include water, argon, nitrogen, and carbon dioxide. Nitrogen and carbon dioxide are by far the most widely used and the easiest to use.

Carbon dioxide is the most versatile of the physical blowing agents as it can be readily made supercritical starting from either the liquid or the gas state. A different pumping system is required to use carbon dioxide in the liquid or the gas state; the advantages and disadvantages of a liquid pump or a gas pump system will be discussed later.

Nitrogen is the least expensive of the physical blowing agents to use but it can only be economically used starting from the gas state. A summary of the properties of nitrogen and carbon dioxide is given in Table 3.1.

Unless required for food, medical, or other potentially regulated uses, microcellular processing requires nitrogen or carbon dioxide of only industrial grade or better purity. Table 3.2 shows the industrial grade specifications for the two gases as set by the Compressed Gas Association (CGA).

Nitrogen is stored as a gas in high pressure cylinders or as a liquid in dewars and tanks. Carbon dioxide is stored as a liquid in high pressure cyclinders, dewars, or tanks. When requesting dewars or tanks from the gas supplier, make sure the provided dewar or tank has a gas use valve. Only use the appropriate gas or liquid outlet to connect to the SCF pump system. Chart Industries' Dura-Cyl dewar is an example of an appropriate dewar; a dewar schematic is shown in Fig. 3.1 as a reference.

Table 3.1 *Properties of Carbon Dioxide and Nitrogen*

Property	Carbon Dioxide	Nitrogen
Chemical formula	CO_2	N_2
Molecular weight, atomic mass units	44	28
Critical point:		
– Temperature, °C	31	–147
– Pressure, MPa	7.37	3.40
Triple point		
– Temperature, °C	–57	
– Pressure, MPa	0.52	
Liquid density @ 21 °C, g/cm^3	0.763	
Gas density @ 21 °C/ 0.10 MPa, g/cm^3	0.00183	0.00116

Table 3.2 *Industrial Grade Specifications for Nitrogen and Carbon Dioxide*

Property	Nitrogen	Carbon Dioxide
CGA type	For gas: G-1.0 type II grade F For liquid: G-1.0 Type I Grade F	For liquid cylinders: G-6.2 grade H type I For liquid dewars: G-6.2 grade H type II
Purity	> 99.997%	> 99.8%
-major contaminants	Water, oxygen	Water, oxygen
Oxygen	< 10 ppm v/v	< 30 ppm v/v
Dewpoint	< – 76 °F	< – 60 °F

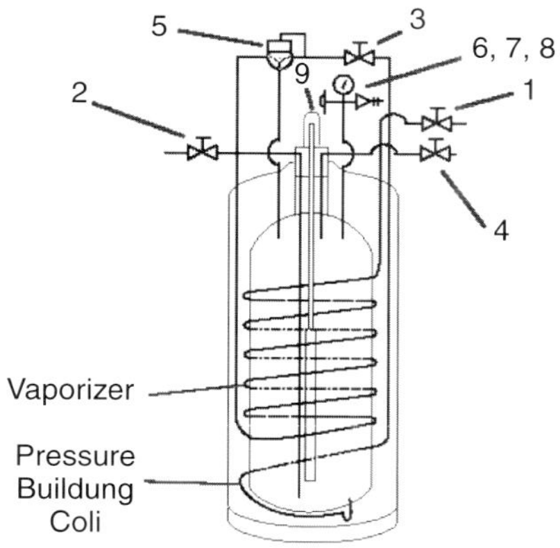

Nomenclature

1. Gas Use Valve – For gas withdrawal.
2. Fill / Liquid Valve – For filling or liquid withdrawal aperations.
3. Pressure Control Valve – To isolate (on/off) the pressure control regulator.
4. Vent Valve – To vent pressure.
5. Combination Pressure Control Regulator – To automatically control operating pressure.
6. Pressure Gauge – Indicates cylinder pressure
7. & 8. Relief Valves, Rupture Disk
9. Liquid Level Gauge – To approximate the liqiud contents of the liquid cylinder.

Figure 3.1 *Schematic of Dura-Cyl dewar (Chart Industries)*

3.3 Nitrogen vs. Carbon Dioxide

Both nitrogen and carbon dioxide are widely used in microcellular processing. However, the choice of blowing agent affects the microcellular structure that is obtained. Therefore, the choice between nitrogen and carbon dioxide should be made depending on what microcellular structure is desired rather than on ease of use or blowing agent costs.

Table 3.3 shows that carbon dioxide generally has a much greater solubility in molten polymers than nitrogen. This indicates that much more carbon dioxide can be added to the molten polymer in microcellular processing than nitrogen. The result of greater blowing agent concentration in the molten polymer is generally greater density reduction than can be accomplished with nitrogen.

***Table 3.3** Estimated Maximum Gas Solubility at 200 °C/27.6 MPa [1-3]*

Polymer	Carbon dioxide (%)	Nitrogen (%)
PE	14	3
PP	11	4
PS	11	2
PMMA	13	1

However, because of the similar diffusion rates of nitrogen and carbon dioxide in molten polymers as shown in Table 3.4, nitrogen tends to generate smaller cells at the same concentration in molten polymer as carbon dioxide. The driving force for nitrogen to devolve from the molten polymer/blowing agent single-phase solution is greater than for carbon dioxide and thus more nucleation sites form with the nitrogen blown molten polymer. Because of the similar diffusion rates, the nucleation sites grow at the same rate with both nitrogen and carbon dioxide and thus nitrogen has smaller cell sizes.

***Table 3.4** Estimated Diffusion Coefficient at 200 °C [1-3]*

Polymer	Carbon Dioxide (cm^2/s)	Nitrogen (cm^2/s)
PS	1.3×10^{-5}	1.5×10^{-5}
PE	2.6×10^{-6}	8.8×10^{-7}
HDPE	2.4×10^{-5}	2.5×10^{-5}
LDPE	1.1×10^{-4}	1.5×10^{-4}
PTFE	7.0×10^{-6}	8.3×10^{-6}
PVC	3.8×10^{-5}	4.3×10^{-5}

3.4 SCF Delivery Equipment

Microcellular processing requires the consistent dosing of supercritical fluid into the molten polymer stream within the extruder or plasticator barrel. For extrusion processing and other plastics processing operations that have a continuously turning screw, a continuous SCF delivery system and blowing agent injectors are needed.

Discontinuous processes, including microcellular molding, require a device to turn the consistent continuous flow from the SCF delivery system into a discontinuous flow. Alternatively, a discontinuous pumping system such as a high-pressure syringe-type pump may be used; at this time, an economically feasible discontinuous pumping system has not been developed and will therefore not be discussed further.

Figure 3.2 schematically represents the SCF delivery system for microcellular molding. A blowing agent source is connected to an SCF pump system (A) which is connected to a bypass manifold (B). The bypass manifold turns the continuous output of the SCF pump system into discrete doses of SCF that are sent to the pneumatic injection valves (C) and through the injectors (D) into the extruder or plasticator barrel.

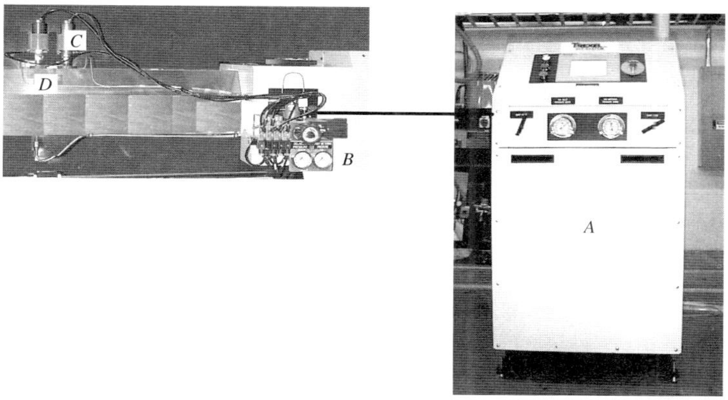

Figure 3.2 Schematic representation of delivery system for microcellular molding

Extrusion SCF delivery systems have a blowing agent source connected to an SCF pump system (A) that is then connected directly to the injectors (D) in the extruder barrel.

3.4.1 Mass Metering vs. Volume Metering

Continuous pumping systems can meter using either volume or mass. Mass metering is much more accurate and consistent for microcellular processing, especially in small or low percent dosage applications. Mass metering requires the additional measurements of temperature and pressure of the volumetric flow and thus pumping systems for mass metering usually additionally include a thermocouple and a pressure transducer.

3.4.2 Liquid Pump Systems

Liquid pump systems capable of making SCF have been developed and used by numerous companies including Trexel and Mitsui. The advantage of liquid pump systems is the ease of liquid mass metering, the simplicity of the delivery system design, and the ready availability of pumps to compress liquid to SCF even at extremely low flow rates. Commercially available diaphragm pumps, such as those by Zenith Products (division of Parker-Hannifin) and other companies, work reliably in this application. The basic schematic of a liquid pump system is shown in Fig. 3.3.

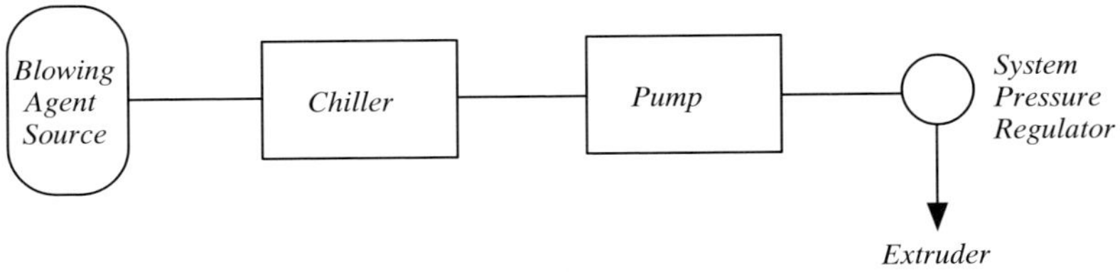

Figure 3.3 *Basic schematic of liquid pump*

Usually mass flow metering devices are not needed for liquid pump systems. Because of built-in pump adjustability, the mass flow output profile of a pump can easily be determined by adjusting stroke length, stroke speed, and head pressure as the variables. Head pressure is adjusted using the system pressure regulator.

Liquid pump systems have two major disadvantages. The first is a requirement for a coolant system for keeping the material that is a gas at room temperature and pressure in the liquid state prior to pressurizing to SCF. The simplicity of the liquid pump system leads to reduced capital costs for a liquid pump system versus a gas pump system but because of the cooling system, the cost to operate a liquid pump system can soon overcome any savings in capital cost. The second disadvantage is the requirement for using a physical blowing agent that can be readily cooled to the liquid state. Carbon dioxide can easily be cooled to the liquid state but nitrogen cannot.

Another disadvantage of liquid pump systems is that satellite metering systems are more difficult to use on the basic liquid pump system design. Satellite metering systems will be discussed further later in this chapter.

3.4.3 Gas Pump Systems

Gas pump systems are by far the most widely used SCF pump systems. Gas pump systems are preferred, especially for microcellular molding, because they work with nitrogen as a blowing agent and because they can readily be used with satellite metering systems.

Figure 3.4 shows a basic schematic for a gas pump system. Unlike the liquid pump system, no chiller exists between the blowing agent source and the pump. The pump is fed directly from the gas outlet of the blowing agent source. High-pressure capable gas pumps are commercially available from Haskel and other companies. Most SCF-capable pumps are two-stage pumps; the first stage pump compresses the gas from about 1.4 MPa pressure to 14 MPa and the second stage pump compresses the output of the first stage pump to a maximum pressure of about 52 MPa.

Gas pump systems require mass flow metering and control for consistent SCF flow. Mass flow metering and control can be accomplished through two main means. The first means is to meter and control the SCF flow with a fixed system pressure and back pressure. A coriolis mass flow meter, such as available from MicroMotion, Liquid Controls, and other companies, work well when followed by an adjustable valve such as is available from

Badger Valve or others. A controller automatically adjusts the valve opening based upon the flow measured by the flow meter.

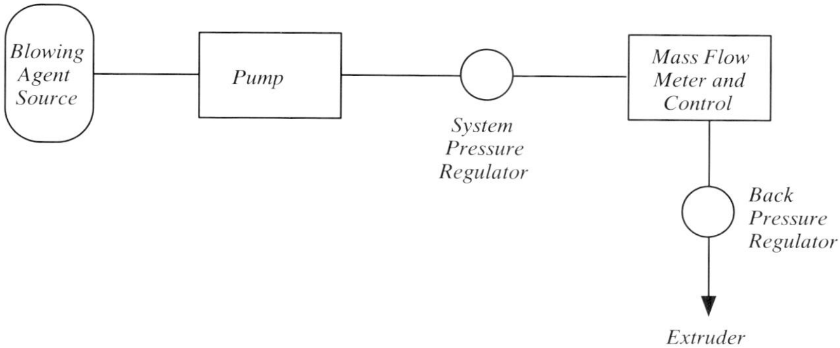

Figure 3.4 *Basic schematic of gas pump*

The second means to mass meter and control SCF flow is to adjust the differential between the system pressure and back pressure across a calibrated opening. A controller reads the system pressure, back pressure, and temperature of the flowing SCF and controls the system pressure regulator, back pressure regulator, or both to adjust the pressure differential across the fixed and calibrated opening to maintain a consistent SCF mass flow.

The latest Trexel SCF pump system (Fig. 3.5) follows this design and offers greater consistency, reduced mechanical hardware, and ease of operation versus the gas pump design using an adjustable metering valve.

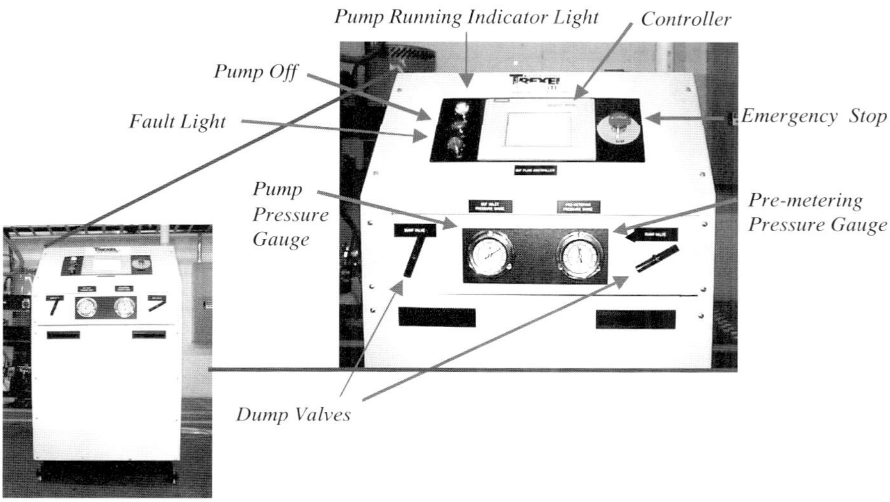

Figure 3.5 *Trexel® SCF pump system*

3.4.4 Satellite Gas Systems

Satellite gas systems may be used in operations running more than one microcellular processing equipment. The satellite gas system is of the same design as a gas pump system except without the most expensive component – the pump. One or more satellite gas systems can be used in conjunction with one gas pump system as shown in Fig. 3.6a or more preferably to prevent shutting down multiple extruders, injection molding machines, or other processing equipment, the satellite gas systems can be used in conjunction with two gas pump systems as shown in Fig. 3.6b.

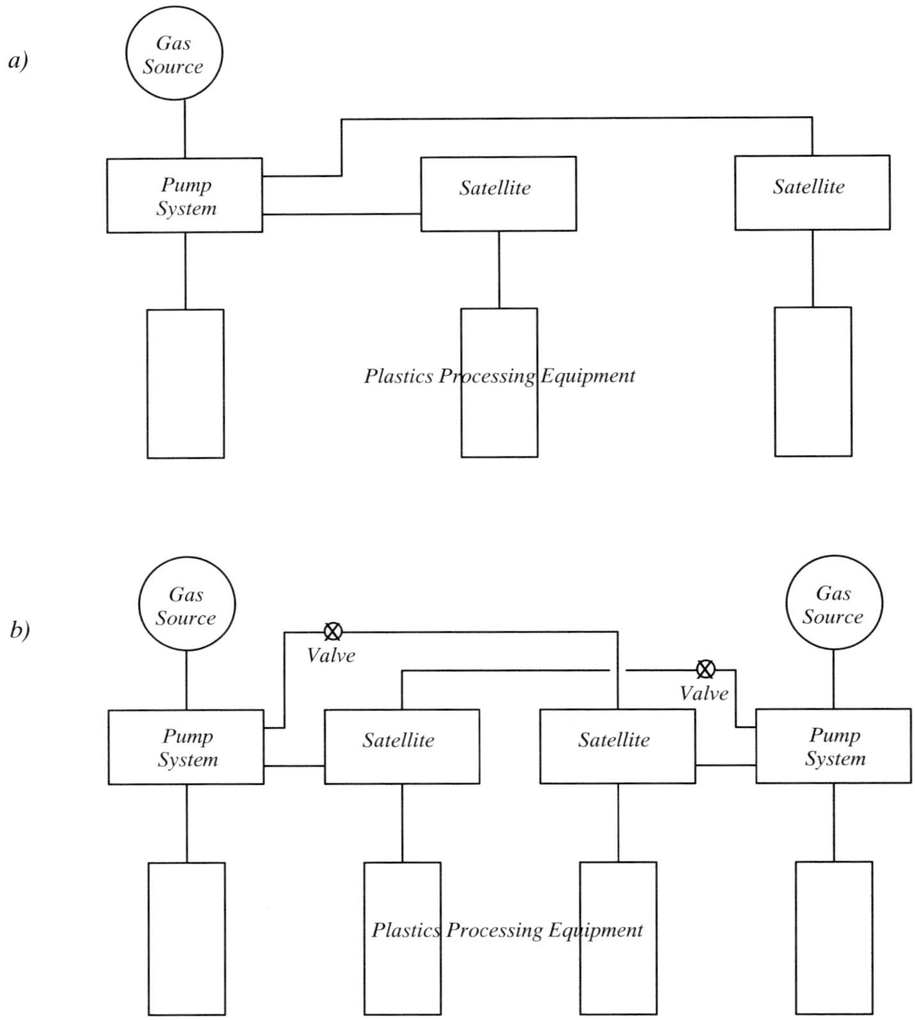

Figure 3.6 *Examples of satellite gas system arrangements*

3.4.5 SCF Injectors

Besides the pump system, the SCF injectors are the other common element of the total SCF delivery system in microcellular processing. Trexel's MuCell SCF injector (Fig. 3.7) may have several tiny holes that help to break up the SCF flow into small streams as it enters the extruder barrel. Low flow or dosage microcellular processing will often have only one tiny hole per injector. Also, the MuCell SCF injector is designed to match the profile of the extruder barrel inner diameter.

Figure 3.7 *MuCell® SCF injector*

The SCF injector is mated with the extruder barrel through a special barrel drilling that allows the injector to seat properly. When the injector seats properly, molten polymer and blowing agent do not leak from the port. An extruder may have one or multiple injectors arranged radially around or linearly along the barrel.

3.4.6 Injector Valves

Two types of injector valves are used in microcellular processing. Extrusion microcellular processing uses simple spring-loaded ball valves that open when the SCF blowing agent flows into the extruder barrel and closes when the SCF does not flow. The closing of the ball valve is necessary to prevent flow of molten polymer into the blowing agent lines when no SCF is being introduced into the barrel. In the MuCell design, the spring-loaded ball valve is designed into an insert that fits into the MuCell SCF injector to put the valve close to the extruder barrel.

Microcellular molding uses actuated injection valves. The valves can be actuated pneumatically, hydraulically, or magnetically. Actuation of the injection valves is required in microcellular molding and other discontinuous processing to control the flow of SCF into the extruder barrel. Opening and closing of the injection valve is actuated by a controller linked to the processing equipment.

Figure 3.8 shows a MuCell® SCF injector valve. The valve is pneumatically controlled for both opening and closing by a controller tied to the molding machine. As in extrusion injection valves, the barrel end of this injection valve has a spring-loaded ball valve to prevent flow of molten polymer into the rest of the injector valve and SCF conduit.

SCF Outlet

Figure 3.8 *MuCell® pneumatic injector valve*

3.4.7 Bypass Manifold

The continuous SCF flow of the pump system is turned into discontinuous flow for MuCell microcellular molding using a bypass manifold. The bypass valve is controlled by a controller tied to the molding machine and it directs the incoming SCF flow to the injection valves and thus to the extruder barrel or to a bypass line for bleeding off the extra blowing agent. With an appropriately designed pump system, the excess blowing agent may be recycled back to the pump.

The MuCell bypass manifold (as shown in Fig. 3.9) also is comprised of the post-metering pressure (or back pressure) regulator discussed above under the gas pump system design, the air actuation controllers for the injection valves and a delivery pressure regulator. The delivery pressure regulator controls the pressure differential between the injected SCF and the barrel pressure at injection. Proper use of the delivery pressure regulator will be discussed in Chapter 5.

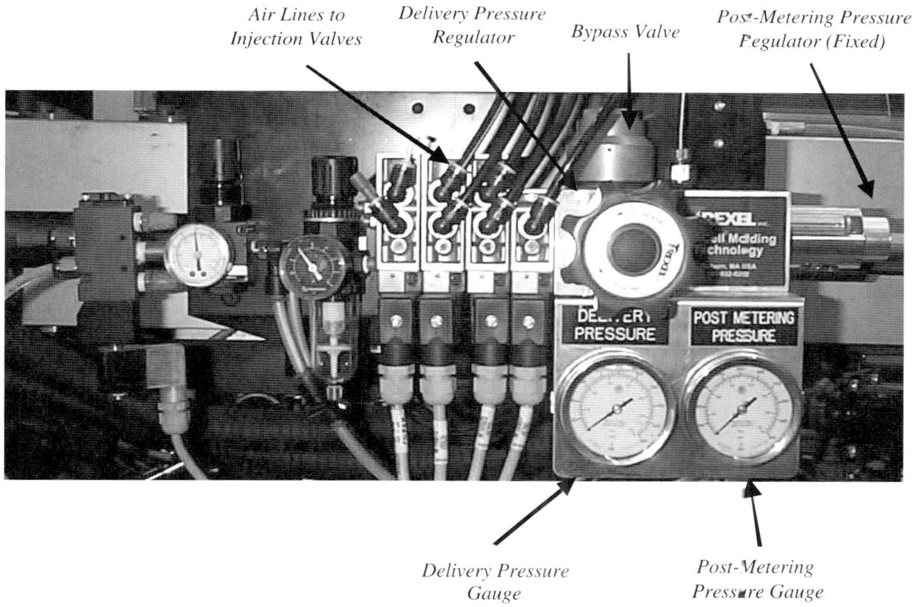

Figure 3.9 *MuCell® bypass manifold*

References

1. Durrill, P.L., Griskey, R.G., *AIChE J.* (1966), p. 1147
2. Durrill, P.L., Griskey, R.G., *AIChE J.* (1969), p. 106
3. Park, C.B., Ph.D. Dissertation, Massachusetts Institute of Technology, Cambridge, MA (1993)

4 Microcellular Molding: The Basics

4.1 Background

Injection molding of microcellular parts is the most widely practiced commercial MuCell® process around the world, despite the fact that extrusion processing had at least a five year head start in development. Development in microcellular molding began by 1996 at Trexel, Inc. with the first injection molding machine modified for microcellular processing operating by mid-1997 in Trexel's development laboratory. This first machine was an Engel 150-ton screw and plunger model but reciprocating screw machines followed soon thereafter.

Commercial introduction of microcellular molding by Trexel occurred at the National Plastics Exposition (NPE) 2000 in Chicago, Illinois, USA; Trexel called the process MuCell® microcellular molding. At the commercial introduction, Trexel also announced that several injection molding machine original equipment manufacturers (OEMs) were licensed to sell new machines with a microcellular processing option.

Evidently, a few other injection molding companies in Japan, Europe, and Korea were also working on developing a microcellular molding process prior to Trexel's announcement. However, these efforts were not as successful in developing a commercially viable process. Thus, most of these companies have either abandoned their efforts or are now Trexel licensees. Table 4.1 lists the Trexel injection molding OEM licensees.

Table 4.1 *Trexel Injection Molding OEM Licensees (as of 1/1/03)*

Arburg, Germany	Italtech, Italy
Battenfeld Kunststoffmaschinen GmbH, Germany	JSW, Japan
Demag Ergotech, Germany	KraussMaffei, Germany
Dongshin Hydraulics Co., Ltd., Republic of Korea	Mitsubishi Heavy Industries, Ltd., Japan
Engel, Germany	Toshiba Machine, Japan
Ferromatik Milacron, United States	Toyo Machinery & Metal, Japan
Husky, Canada	Van Dorn Demag, United States

At least two other organizations have since announced that they are working on alternative microcellular molding technology to Trexel's design. These designs will be discussed further in Chapter 11 of this book.

4.2 Advantages of Microcellular Molding

Microcellular foam injection molding is a process and equipment technology for plastic injection molded components that results in an average cost reduction of 16 to 20% compared to conventional molding net of all equipment and licensing related costs.

The microcellular molding process changes the fundamental cost structure of injection molded parts through four main sources of economic benefit:

1. Reduced operating costs through cycle time reductions of up to 50%, reduced scrap rates, and lower energy consumption;
2. Lower capital costs through the purchase of smaller and fewer machines, and fewer and less expensive molds;
3. Reduced material costs through component density reduction, thinner design, and material substitution, and
4. The ability to mold thermoplastic parts that are flatter, straighter, and dimensionally improved due to the elimination of the molded-in stress associated with the pack and hold phases of conventional injection molding.

Greater detail of the microcellular processing advantages will be given in later chapters along with several case studies.

4.3 Comparison to Other Molding Processes

A number of older molding processes also use gas or a blowing agent. These processes include:

- Structural foam molding
- Gas assist molding and
- Foaming with chemical blowing agents (CBAs).

How is microcellular molding different from these processes?

The two main differences between microcellular molding and the three alternative processes are:

1. Microcellular molding works for parts of any thickness and is able to mold thin parts (less than 3-4 mm thick) where none of the other processes have been particularly effective.
2. Microcellular molding facilitates a more controlled weight reduction and generally homogeneous cell distribution throughout the part.

However, microcellular molding is not intended to compete with these other approaches though its usefulness may overlap with that for structural molding, gas assist molding, and CBA blown injection molded parts.

4.3.1 Microcellular Versus Structural Foam Molding

Structural foam molding is generally used for big parts and, most commonly, with HDPE on specialized low-pressure molding machines. Weight reduction over solid molding can be 10% or greater. Microcellular molding can be far superior to structural foam processes in terms of material reduction and cycle time reduction for most materials, including all of the commonly used engineering materials. Microcellular molding can also fill very thin sections at the same time as the thick sections providing greater design flexibility. However, microcellular molding does not provide any major advantages over existing techniques with high length-to-thickness parts and with thick (greater than 6 mm thick) parts, especially in HDPE.

4.3.2 Microcellular Versus Gas Assist Molding

Gas assist molding provides Class A surfaces by running a void through the middle of a relatively thick part using special mold and part designs. As stated above, the microcellular molding process does not work well in thick parts. Additionally, microcellular molding should not be expected to provide a Class A surface.

However, gas assist molding is often used merely to eliminate sink marks. In this case, microcellular molding may be a better alternative as microcellular molding generally produces higher weight reduction, faster cycle times, and less warpage than gas assist molding while still eliminating sink marks.

4.3.3 Microcellular Versus Chemical Blowing Agents

Chemical blowing agents (CBAs) are designed to decompose at a specific temperature, yielding the gaseous blowing agent. Various grades are designed for a range of decomposition temperatures to yield the blowing agent as close to the mold cavities as possible. CBAs are often used in thick parts to eliminate sink marks while at the same time producing density reductions as well. The use of CBAs in thin parts usually produces very poor surfaces and considerable porosity in the parts thus dramatically reducing the physical properties of the molded part. Also, higher density reductions are usually not economically feasible with CBAs.

Other issues problematic when using CBAs that microcellular molding overcomes include:

- Many endothermic CBAs generate water (as well as carbon dioxide blowing agent), which necessitates additional water scavenger additives to prevent degradation of the molten polymer due to water.
- CBA batch-to-batch inconsistency of chemical blowing agents that lead to having to continuously adjust process conditions.

- CBA's inherent thermal instability that makes CBAs more difficult to use in high-temperature resins.
- CBAs generate residual byproducts that remain in the plastic or the CBAs may not completely decompose to generate the gaseous blowing agent. The byproduct and non-decomposed CBA may lead to poor organoleptic properties, mold vent plugging, and to difficulty for use in in-house scrap recycling programs.

4.4 Microcellular Molding Equipment Basics

Microcellular molding equipment has two primary components: the injection molding equipment itself and a supercritical fluid (SCF) delivery system. The SCF delivery system is described in detail in Chapter 3.

The injection molding equipment is basically standard equipment; however, several modifications are made to facilitate the microcellular process. The changes are summarized below and illustrated in Fig. 4.1.

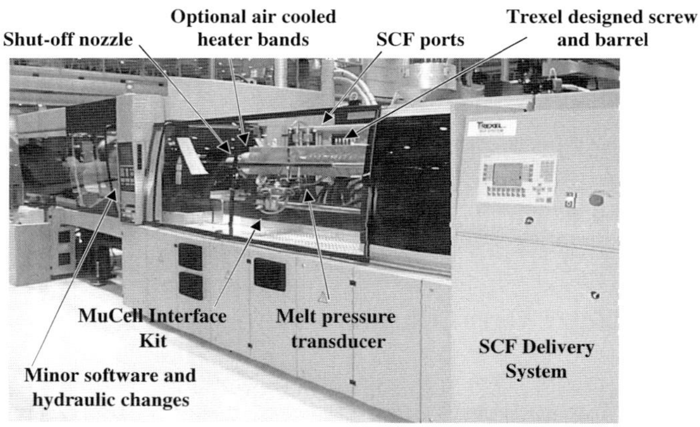

Figure 4.1 *Injection molding machine modifications to run microcellular process*

- The barrel must include drillings for SCF injection ports, pressure transducers, and rupture discs.
- A shut-off nozzle is added.
- A specially designed microcellular processing screw is installed.
- On new machines, programming changes are made to the machine controller.
- Optionally, the barrel length to diameter (L:D) ratio is increased.
- Optionally, air-cooled heating bands may be added to the end of the barrel.

Even with the machine modifications, the microcellular process capable injection molding machine can still run both solid molding and microcellular molding.

4.4.1 Shut-off Nozzle

The shut-off nozzle is a critical component for making the microcellular process work. The shut-off nozzle must be closed during all phases of the injection molding cycle except for injection and hold. To ensure this, the nozzle should be positively actuated to close. The shut-off nozzle is kept closed as 1) an open nozzle would allow the SCF laden molten polymer in front of the screw tip to foam and drool out of the opening and 2) the closed nozzle allows a constant and consistent back pressure to be maintained on the molten polymer. When the shut-off nozzle is closed, the back pressure can be programmed to maintain a pressure high enough to prevent the SCF from coming out of the SCF/molten polymer solution in the barrel. It is important that the shut-off nozzle is designed and manufactured according to the necessary pressure maintenance requirement.

Shut-off nozzles are readily available from most injection molding OEM manufacturers or some parts suppliers, including Herzog AG (Degersheim, Switzerland).

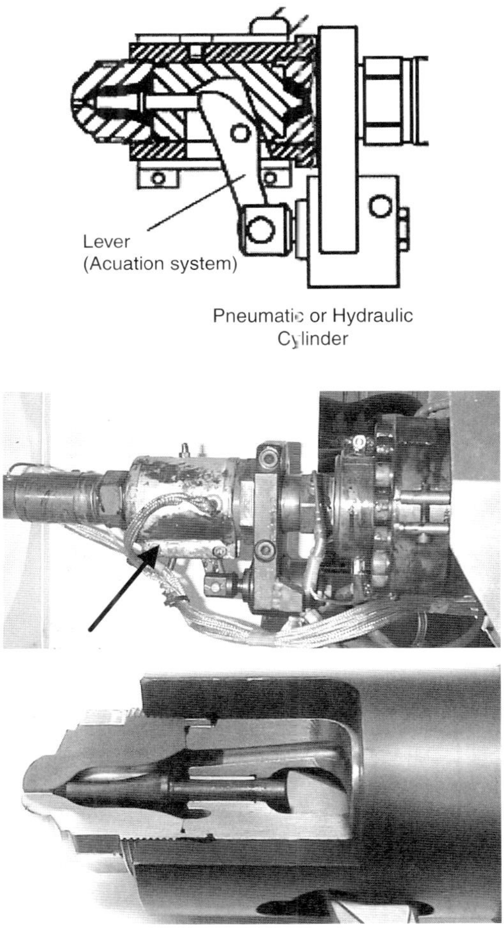

Figure 4.2 *Shut-off nozzle: Herzog needle-type design*
Top: schematic (from Herzog AG web site)
Middle: shut-off nozzle installed
Bottom: cutaway showing internal detail

4.4.2 Screw Design

A standard screw for injection molding must be able to melt, meter, and inject plastic into a mold. A screw designed for the microcellular process must also meet the following additional requirements.

- The screw must hold constant and consistent pressure during all phases of the injection molding cycle, except during injection.
- The screw must prevent escape of the injected SCF out of the feed throat.
- The screw must be able to mix the injected SCF into the molten polymer to form a consistent single-phase solution

To meet the standard requirements, a microcellular screw has conventional feed and compression zones with a short metering zone for the first 16D to 20D (D is the inner diameter of the screw barrel and the length is measured starting from the center of the feed port).

In a microcellular processing screw, the short metering zone is followed by a pressure restriction element. The purpose of this element is to reduce or eliminate reverse flow of the SCF blowing agent in the screw and barrel; the results of reverse flow of SCF would be blowing agent coming out of the feed throat as a small steady stream of gas or as small pops. The pressure restriction element also helps to maintain a constant and consistent pressure downstream.

The pressure restriction element can be of numerous design types including reverse flight elements, sliding check ring, or ball check valves. Figure 4.3 shows an example of a sliding check ring design. The design could also incorporate a spring for faster actuation.

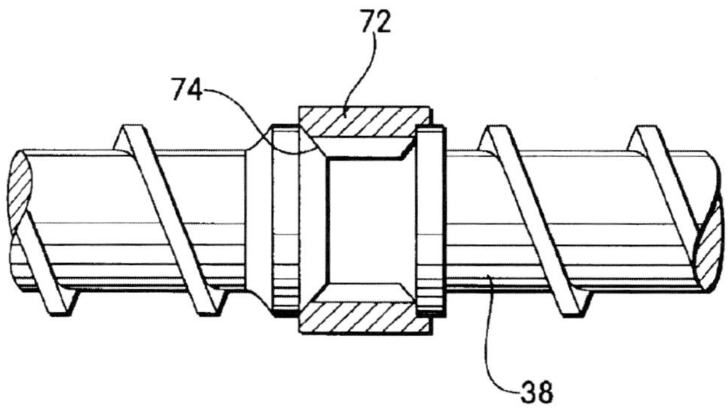

Figure 4.3 *Pressure restriction element – check ring design [from U.S. Patent 6 322 347]*

Figure 4.4 shows how a sliding check ring would operate when open and closed. In both diagrams, the feed throat would be to the left and the nozzle to the right. Molten polymer normally flows from left to right or in a downstream direction. When the ring valve is open, the molten polymer can flow through and past the restriction element; this would be the normal valve position when the screw is reciprocating and building up material in front of the screw for a shot. However, when the screw moves forward to inject the SCF/molten polymer mixture, the sliding check valve closes to block the backward flow of the SCF/ molten polymer solution as indicated by the arrow in the lower diagram.

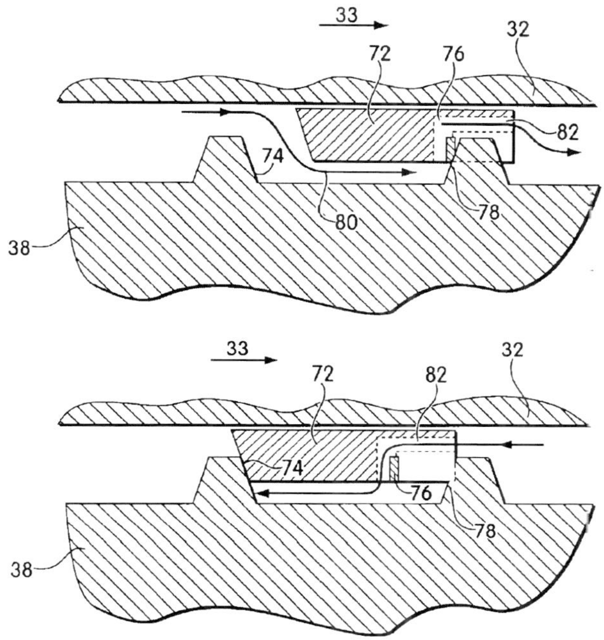

Figure 4.4 Check ring action when open (top) and closed (bottom) [from U.S. Patent 6322347]

An example of a ball check valve is shown in Fig. 4.5. The left photo shows the entrance and exit holes to the ball check that allow molten polymer to pass by the melt dam on the screw. The right photo shows some of the flow channel through the screw, including the ball valve itself.

Following the pressure-restriction element on a microcellular processing screw is a short wiping section. This section usually is only 1 to 4D long. The purpose of this section is to break the injected SCF blowing agent into smaller discrete masses for easier dissolution in the molten polymer as schematically shown in Fig. 4.6. This section may consist of broken screw flights but usually has continuous flights. To help maintain pressure on the molten polymer, the wiping section usually has little to no decompression compared to the upstream metering section.

Following the wiping section is a mixing section (see Fig. 4.7). The purpose of this screw section is to evenly disperse the SCF blowing agent throughout the molten polymer. The mixing forms the SCF/molten polymer single-phase solution that is critical for making consistent microcellular molded articles. The mixing section design can be any of a wide number of designs in the plastic industry including broken, cut or slotted screw flights, or Maddock-type mixers.

Figure 4.5 *Pressure restriction element – ball check valve*

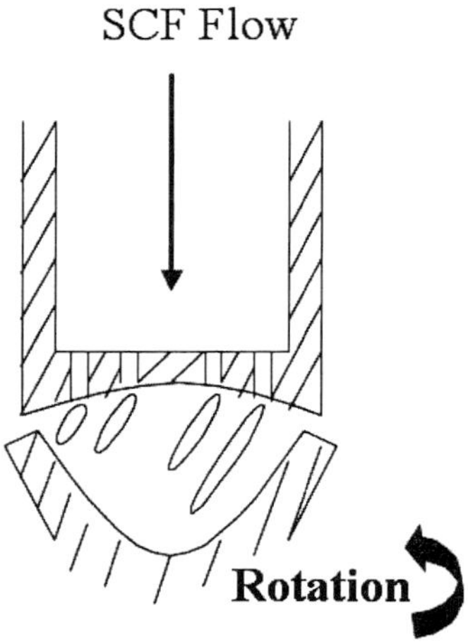

Figure 4.6 *Schematic of wiping action against injector*

The end of the microcellular processing screw has a screw tip valve. This valve is standard in the injection molding process to prevent upstream flow of molten polymer when the screw travels forward during injection. The tip valve plays the same role in microcellular processing; however, the tip valves are designed to provide improved closing consistency and speed.

To fit all the described zones onto the screw, the screws and barrels on the new OEM injection molding machines had a 26:1 or 28:1 L:D length. However in November 2002, Trexel announced that new and modified injection molding machines for the MuCell microcellular process will have screws and barrels of standard lengths, i.e., 22:1 to 24:1 L:D.

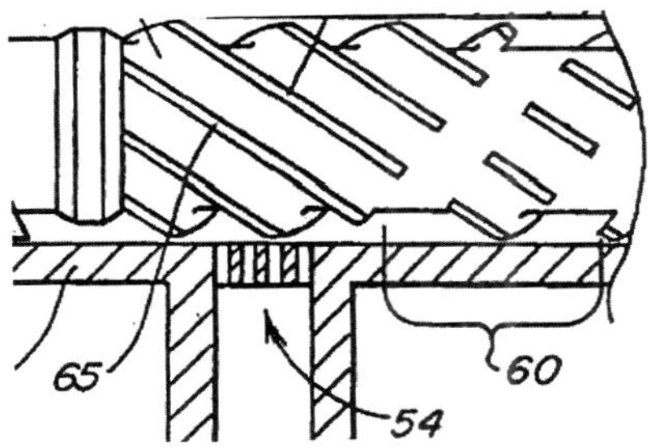

Figure 4.7 *Schematic of wiping and mixing sections [from U.S. Patent 6 284 810]*

4.4.3 Barrel Modifications

Numerous modifications associated with the extruder barrel may be made for microcellular processing. Some of the modifications are necessary while others are optional. The necessary changes include tapping the barrel for SCF injectors, a rupture disc, and a pressure transducer. Also, the barrel may need an extension or may need to be replaced if the screw is of a different length than the original design. The optional changes include tapping the barrel for an additional pressure transducer and adding air-cooled heater bands.

The microcellular molding machine may have one or more SCF injectors threaded into the barrel; usually the MuCell process uses two injectors. The multiple injectors are used to allow the addition of SCF blowing agent into the molten polymer at the wiping section of the screw throughout reciprocation. The downstream port is located near the furthest forward screw position and the upstream port is located at an adequate distance from the downstream one. With two injectors, the downstream injector operates first during screw reciprocation. However, as the screw continues to move backwards, the downstream injector may no longer be over the wiping section of the screw; when this occurs, the downstream injector would no longer operate and the upstream injector operates. The injectors are operated by a properly modified molding machine controller or a controller specific to the microcellular process.

A rupture disc is installed in the barrel for safety. The rupture disc is usually located on the non-operator side of the molding machine near the position of the downstream injection port.

The required pressure transducer is also located near the position of the downstream injection port. This pressure transducer would measure the melt pressure in the barrel of the molten polymer near where the SCF blowing agent is being injected. This pressure is called the SCF inlet melt pressure (SCF IMP) and is critical for maintaining consistent SCF addition.

The optional pressure transducer can be located near the downstream end of the barrel. This pressure transducer measures the molten polymer pressure downstream of the screw and gives a direct measurement of back pressure in a solid molding process or microcellular process pressure (MPP) in a microcellular process.

The optional air-cooled heater bands may be used downstream of the injection ports. The microcellular process uses lower barrel and melt temperatures than in the solid process at positions after the introduction of the SCF blowing agent; air-cooled heater bands help to cool the barrel faster and to obtain a consistent process on start-up of the microcellular process for manufacturing.

4.4.4 Control Modifications

New injection molding machines for microcellular processing often have a specific page in the injection molding machine controller for inputting and following parameters specific to the microcellular molding process. Upgraded or retrofitted equipment requires an additional controller either associated with the SCF delivery system or as a separate item.

An example of a MuCell process specific page offered by an injection molding machine original equipment manufacturer can be seen in Fig. 4.8. The screen shot shows the many added control functions. These functions include:

- Enabling/disabling the microcellular process
- Enabling/disabling the SCF injectors
- SCF injector opening and closing parameters
- Microcellular Process Pressure (MPP) setting
- SCF inlet pressure cutoff setting
- Barrel pressure readings and
- A SCF dosage calculator.

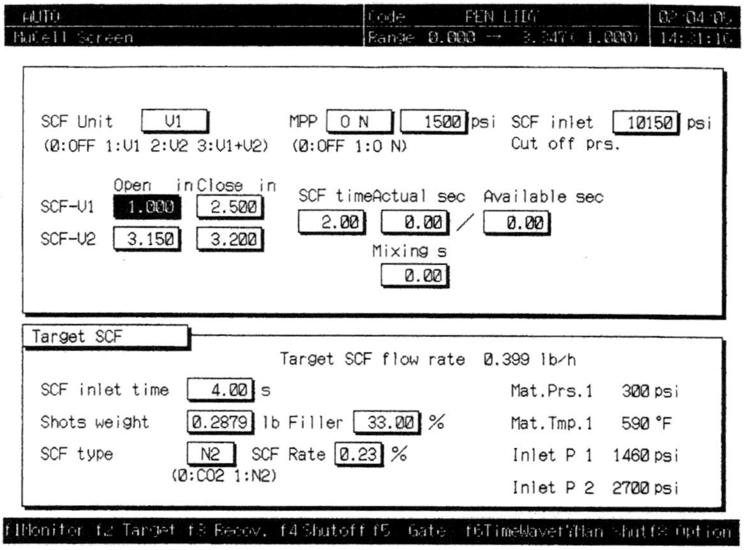

Figure 4.8 Screen shot of JSW molding machine MuCell process page

Further explanation of the use of these parameters will be given in the next chapter on microcellular processing.

4.5 SCF Delivery Equipment

Microcellular molding requires consistent dosing of supercritical fluid into the molten polymer stream within the plastication barrel. The standard SCF delivery equipment used in MuCell microcellular molding includes a gas source, gas pump system, bypass manifold, injection valves, and SCF injectors. The design of SCF delivery equipment for microcellular molding is discussed in Chapter 3.

5 Microcellular Molding: Processing Basics

5.1 Introduction

Microcellular molding provides many benefits to the parts manufacturer and original equipment manufacturer (OEM). The benefits can be divided into two categories: process benefits that reduce processing costs and molded part benefits that improve part functionality or properties. Process benefits include clamp tonnage reduction, cycle time reduction, and weight reduction. Molded part benefits include dimensional stability and improved toughness. However, like any process, all the possible process and part benefits cannot be accomplished simultaneously. A processor must decide what specific improvements are desired from the microcellular molding process in a solid molded part.

Only Trexel's MuCell microcellular molding process will be discussed in this chapter; at this time, there are no other publicly known thermoplastic microcellular molding processes in commercial use. This chapter summarizes the basic procedure and parameters for starting up the microcellular molding process and how to optimize the process for a specific result.

5.2 Theoretical Basis for Microcellular Processing

To better understand the basic process parameters of MuCell microcellular molding, a short theoretical description of the process follows. A more in-depth discussion of the theory can be found in Chapters 2 and 7 and in many of the references listed in Appendix 4.

The MuCell process can be broken into four stages: creation of a single-phase solution, homogeneous nucleation, cell growth, and part formation.

A single-phase solution as defined by Trexel is a molten polymer into which a physical blowing agent (typically either nitrogen or carbon dioxide) in a supercritical state is uniformly dispersed and dissolved. During the injection molding process, this single-phase solution is created by injecting the supercritical fluid (SCF) into the barrel of the injection molding machine during the screw recovery step. As the mixture of SCF and molten polymer is conveyed down the screw, the SCF dissolves into the molten polymer creating the single-phase solution. This single-phase solution has a significantly lower melt viscosity than the polymer itself. Once created, the single-phase solution is maintained by keeping the solution under pressure. The single-phase solution is a necessary requirement for homogeneous nucleation to occur.

From a theoretical perspective, homogeneous nucleation occurs when the single-phase solution undergoes a thermodynamic instability large enough to simultaneously nucleate millions of cell nuclei. With respect to the injection molding process, this instability occurs when the molten polymer under pressure in the injection molding machine's plasticator barrel is injected into the cavity of a mold, which is at a much lower pressure. While it is not possible to instantaneously fill the cavity, injection time is usually sufficiently short to create uniform cell nucleation.

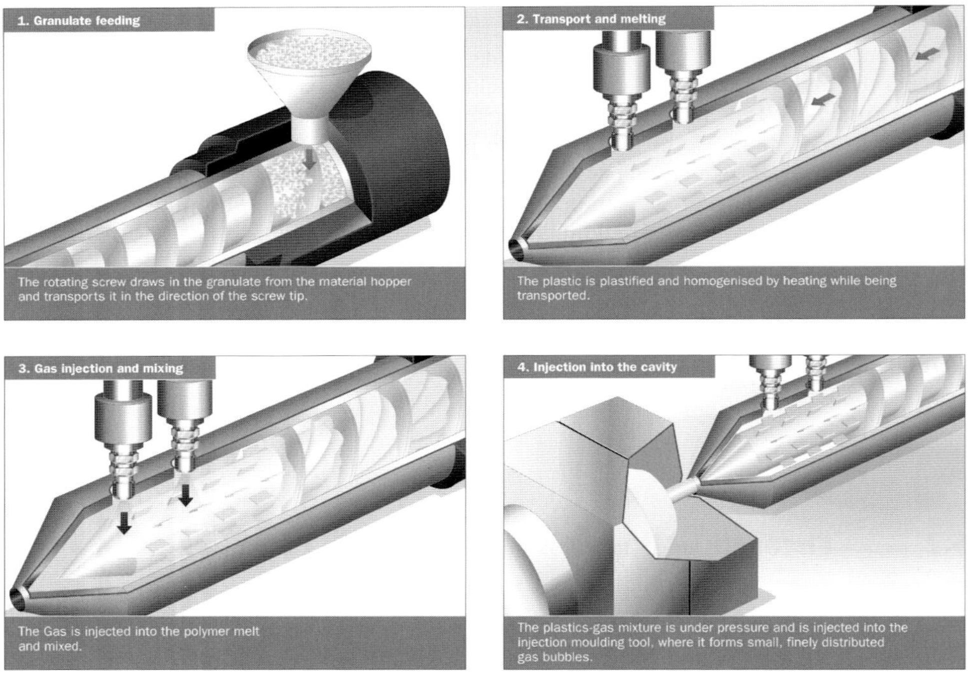

Figure 5.1 *Schematic of microcellular molding process*

Once nucleation is completed, cell growth will start. The cells will grow until the material completely fills the mold, the material develops enough rigidity to withstand the pressure of the gas in the cells, or there is no longer sufficient gas to continue expanding the cells. Cell growth is a function of the restraining forces on the expansion of the cells and the gas pressure in the cells.

The final phase of the process is part shaping. During the injection molding process, the mold provides the means for shaping and cooling the part. The cell size average and cell size uniformity is dictated by material, process conditions, and the mold cooling rate.

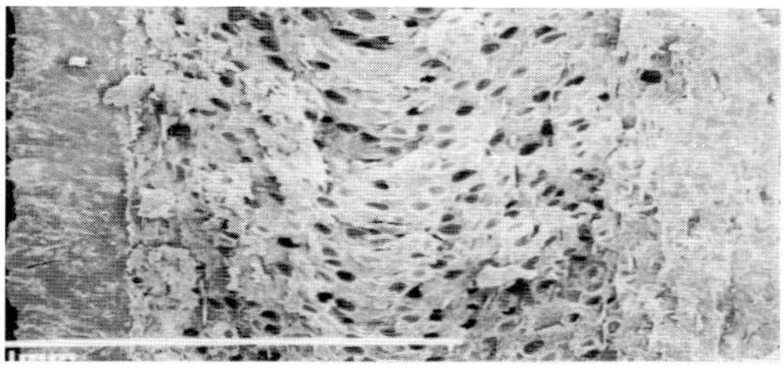

Figure 5.2 Cross-section of microcellular molded part (average cell size = 10 microns)

All aspects of the microcellular molding process are interdependent. For example, when processing with polypropylene, microcellular molding injection speeds are faster than for conventional molding. The increased injection speeds result in higher cell nucleation density, which yields improved cell structure across the part. In turn, the improved cell structure improves part performance as a uniformly small cell structure results in a better retention of properties.

5.3 Process Effects

The introduction of SCF into a polymeric material has a distinguishable effect on the material's properties. In this section, the effects of the MuCell process are reviewed.

5.3.1 Effect on Viscosity

In general, the addition of SCF into a molten polymer reduces the viscosity due to a reduction in T_g, and the reduction of the viscosity follows classical viscoelastic scaling theory [1]. The reduced viscosity due to SCF dissolution in the molten polymer can result in hydraulic injection pressure reductions as compared to the conventional molding. For example, it has been demonstrated that with equivalent material, mold temperatures, and injection velocity, the hydraulic pressure required to fill the mold cavity with a wall thickness of 0.22 mm and a flow length to part thickness ratio of 280:1 was reduced from 130 bar in solid molding to 113 bar using 2% nitrogen as a blowing agent and to 103 bar with 5% carbon dioxide.

The effect of SCF on polymer viscosity depends on the material type, filler type, and filler level. Filler type and level become important in that SCF does not dissolve into the filler and cannot change the viscosity of the filler. Therefore, the apparent change in viscosity for a filled material is not as large as for an unfilled material. In addition, the viscosity benefit for polyolefins is much less evident in thicker wall (greater than 1.5 mm) parts as these materials are normally very low in viscosity.

Figure 5.3 shows the effect of viscosity in ABS as a function of carbon dioxide concentration and shear rate at 188 °C. The viscosity change is especially pronounced at low shear rates. The effect of nitrogen or carbon dioxide dissolution on molten polymer viscosity is qualitatively similar for most resins.

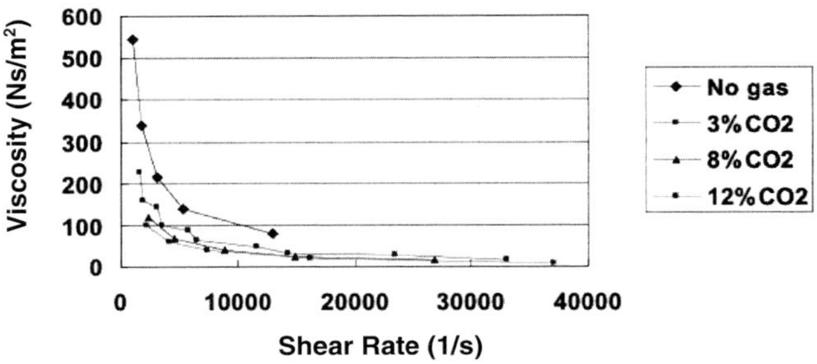

Figure 5.3 *Effect of carbon dioxide on ABS viscosity*

5.3.2 Effect on Part Weight

The MuCell injection molding process does not reduce part weight as much as might be expected from batch process experiments.

In the MuCell process, weight reduction is driven by one primary factor – the ratio of flow length to part thickness (L/t). It can be generally assumed that for an application with a flow length to part thickness ratio of 100:1, the weight reduction will not exceed 15% and as that ratio approaches 150:1, 10% is a more realistic number. Other factors affecting this result include the flow balance of the part, the overall wall thickness, and the filler type and level. Figure 5.4 represents the typical relationship between percent weight reduction and L/t. With this relationship in mind, processors can determine weight reduction expectations for an application molded with the MuCell process.

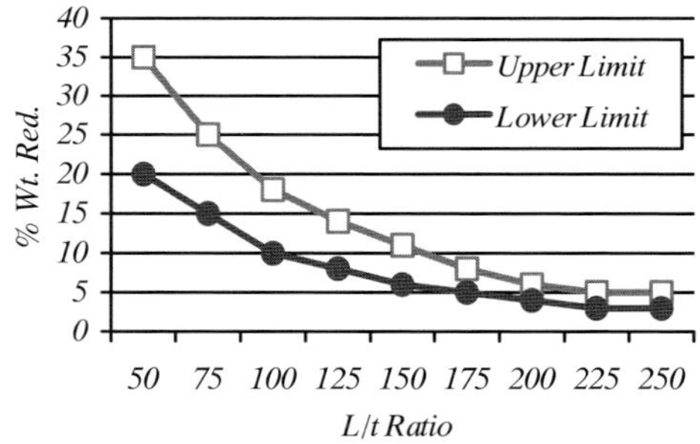

Figure 5.4 *Typical relationship of % weight reduction with L/t ratio*

5.3.3 Effect on Cycle Time

The MuCell process generally tends to shorten cycle time. This occurs primarily due to the elimination of hold time. The uniform internal gas pressure of the microscopic cells provides the cavity pressure needed for filling. As depicted in Fig. 5.5, the pack and hold molding step, a prominent contributor to cycle time and a source of molded-in stress in conventional injection molding, is therefore greatly reduced or eliminated with the micro-cellular molding process. Cavity pressure reductions of up to 80% are achieved with microcellular molding – a particular benefit for insert molding, in-mold labeling, in-mold decorating, and encapsulation, where high internal cavity pressures can lead to deformation and bleed-through.

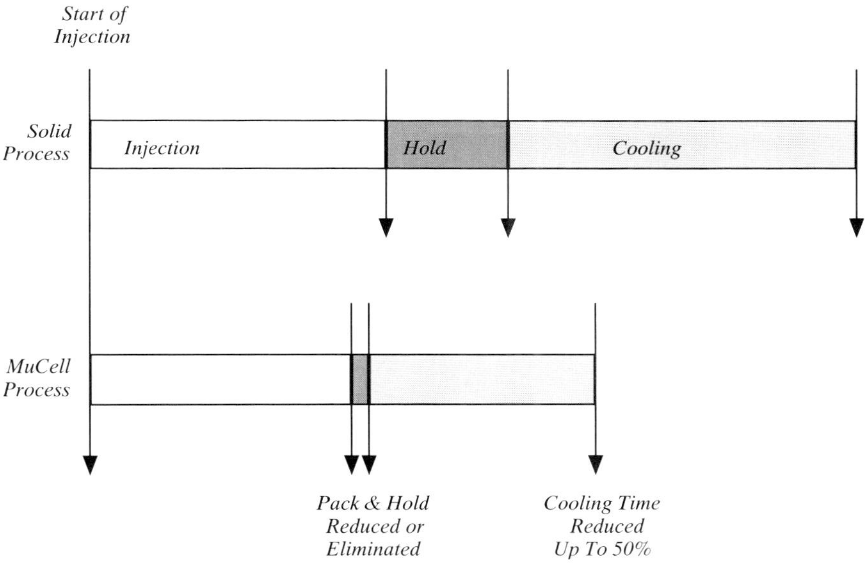

Figure 5.5 *Comparison of solid and MuCell process cycle times*

While the hold phase of the injection molding process is usually eliminated with the MuCell process, some cycle time may occasionally need to be added back to the cooling time for glass-reinforced resins. All added, 20% cycle time reductions are normal for a properly cooled mold. Cycle time reduction is more significant for fiber reinforced materials than for unfilled materials because filled materials develop sufficient modulus to resist cell growth quicker than unfilled materials. Often, the determining factor for cycle time reduction is the quantity and uniformity of cooling in the mold cavity; the hottest location in the part will dictate the maximum cycle time reduction achievable.

Another critical factor in determining the cycle time is part thickness. It is more difficult to significantly reduce cycle times with parts having a thickness greater than 4.0 mm (2.5 mm in PP) because the cells appear to act as a thermal insulator, hence slowing down the rate of cooling. This limitation in cycle time reduction is particularly likely in poorly cooled molds. Excessive part thickness and poor mold cooling can lead to a blowout, as the cells continue to grow in the still molten core of the molded part.

5.3.4 Effect on Temperature Settings

In microcellular molding, the barrel temperature profile in the injection molding machine is maintained as set for a solid molding process up to the SCF injectors. After the injectors, the temperature profile is usually flat. After the injectors, the temperature setting is usually the same as the last zone prior to the SCF injector. For high-temperature resins or when high blowing agent levels are used, the temperature setting after the injectors may be reduced 10 to 25 °C from the last zone prior to the injector.

Mold temperature settings are usually lower than those used for solid molding. However, the amount of mold temperature reduction is dependent on the goals of implementing the microcellular molding process.

5.3.5 Effect on Clamp Tonnage

The microcellular molding process can and usually does drastically reduce the clamp tonnage required to mold a part with the conventional solid process. Clamp tonnage reductions can be as high as 50% or even more. This phenomenon is usually attributed to the drastically reduced resin viscosity at low shear rates and the elimination of the pack and hold phase.

5.4 Beginning Work with the Microcellular Molding Process

Before running a mold using the MuCell microcellular molding process, the mold should be run using the conventional solid molding process first. Running the solid process first provides:

- A baseline for comparing the microcellular parts;
- The startup conditions needed for getting the microcellular process running;
- An opportunity to resolve mold or part design issues; and
- The total solid part weight including runners and sprues.

Once a part is running in solid, an initial SCF blowing agent flow rate setting for the SCF delivery system must be calculated. For a mold never run in the microcellular process before,

- Assume a 10% weight reduction (unless you know that you will be running at some value lower than that);
- Assume an SCF dosage time of no more than 25% of the screw recovery time; and
- Assume an initial SCF level of 0.1% nitrogen or 0.5% carbon dioxide for a filled resin or 0.3% nitrogen or 0.8% carbon dioxide for an unfilled resin unless previous experience with the material indicates otherwise.

Based on the assumptions, use Eq. 5.1 to solve for the initial SCF flow rate.

$$\%\ \text{SCF} = \frac{(\text{SCF Flow Rate})(\text{Valve Open Time})(27.8)}{(\text{Total Shot Weight})} \tag{5.1}$$

Where: SCF Flow Rate in kg/h
Valve Open Time in s
Shot Weight in g
27.8 is a conversion factor

At this time, start up the SCF delivery system, set the flow rate for the calculated initial setting, and allow the flow rate to stabilize. Please see the manual from the SCF delivery system manufacturer for the installation and startup procedure for a specific SCF delivery system. To obtain the proper initial flow rate, the flow element or valve may need to be changed.

Next, set the Microcellular Process Pressure (MPP) on either the molding machine or SCF delivery system controller. The MPP is the actual or calculated pressure maintained on the molten polymer/SCF single-phase solution in front of the plastication screw and is needed to keep the SCF in solution. Thus, as a rule of thumb, the MPP should be at least 70 bar; a good initial setting is 140 bar. On rare occasions, it may be necessary to increase MPP as high as 245 bar; however lowering the MPP helps reduce screw recovery time. The MPP is similar to the back pressure in conventional solid molding but it is maintained throughout screw recovery and idling.

For SCF delivery systems with manual delivery pressure control, set the delivery pressure using the proper regulator to 3 to 14 bar above the barrel pressure measured at the SCF injectors during solid cycling; this pressure is also known as SCF Inlet Melt Pressure (SCF IMP) and then enable the SCF injectors on the molding machine or the SCF delivery system controller.

Next, set the SCF injector-open position to 10 mm screw recovery position. Allowing the screw to begin recovery prior to opening the injector lets the MPP and SCF IMP stabilize first. If the delivery pressure drops more than 14 bar on opening the injector, the injector-open position may need to be increased. If a second SCF injector is needed, the second injector-open position should be set to 2.00D (D = screw diameter).

Set the SCF injector-open time to the initial assumption of 25% of screw recovery time.

The proper delivery pressure, SCF injector-open position, and time can be determined by watching the delivery pressure and SCF IMP during screw recovery. As mentioned previously, the delivery pressure should decrease no more than 14 bar when the SCF injector opens, hold constant pressure during dosing, and then return to the delivery pressure setting when the injector closes as shown in Fig. 5.6.

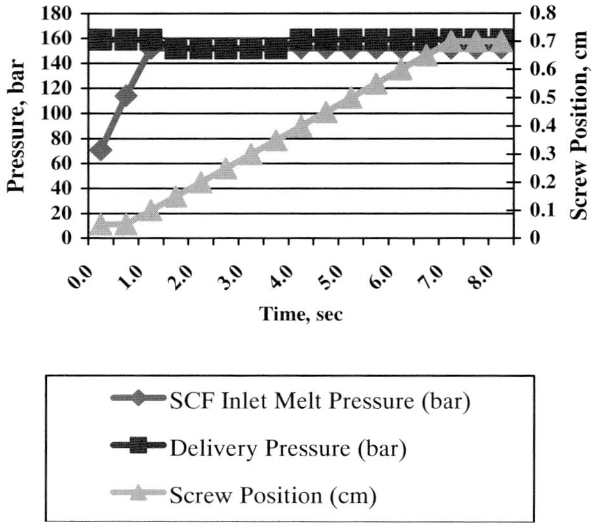

Figure 5.6 *Proper delivery pressure profile*

If the delivery pressure profile does not appear similar to that in Fig. 5.6, the delivery pressure may be set too high or too low, the injector-open time may be too early or the injector-open time may be too long. Each incorrect setting has a corresponding delivery pressure profile that can be used as a diagnostic tool. A combination of incorrect settings may also exist. Observing the delivery pressure profile can also help diagnose a plugged injector.

Set pack and hold time to zero though it may be necessary to maintain some pack and hold time during startup of the microcellular process.

Now everything is set to run the microcellular molding process. The time required for a foam structure to appear in the molded part will depend on the shot size as compared to the barrel volume, but it will usually happen in less than 10 shots. The process should stabilize within a few shots after microcellular foam is seen.

Having now obtained microcellular parts, it is good practice to recalculate the actual SCF % in the molded microcellular part. Use Eq. 5.1 and the actual total part weight, SCF flow rate, and injector-open time.

5.5 Optimizing the Microcellular Molding Process

Once the microcellular molding process has stabilized, the process can be optimized for the desired part characteristics. These include:

- Part weight
- Cycle time
- Cell structure
- Surface appearance
- Clamp tonnage
- Part warpage

It should be emphasized that the microcellular molding process can be optimized for one or several of these properties or molding process parameters for a given molded part, but it cannot be optimized for all. It is important for a processor to prioritize what benefits are most desired from using the microcellular process for a specific molded part.

Process conditions as well as mold and part design can be used to optimize the microcellular process. A more in-depth discussion of mold and part design for the MuCell microcellular process can be found in Chapter 9.

5.5.1 Guidelines

Any of the conventional molding process conditions may be used to optimize the microcellular process including shot size, mold temperature, and barrel temperature settings. However, the microcellular process has additional parameters that may be adjusted. These include: % SCF, injector-open position, injector-open time, MPP, and delivery pressure. The guidelines for setting MPP, injector-open position, and delivery pressure were discussed earlier.

As a general rule, the maximum percent SCF that can be dissolved into molten polymer is 1.5 to 2.0% nitrogen or 5 to 7% carbon dioxide. As the SCF concentration increases, the MPP may need to be increased to keep the SCF in solution.

Most microcellular molding is performed using nitrogen as the blowing agent. As discussed previously, nitrogen tends to make smaller cells of a more uniform size than carbon dioxide. Table 5.1 shows the typical nitrogen levels for various resins; the typical carbon dioxide levels are about three times the levels given for nitrogen. Generally, no more SCF should be used than the amount determined to provide the desired benefits.

Injector-open time is initially set for 25% of the screw recovery. As the microcellular process is optimized, the injector-open time may be varied from 20 to 70% of the screw recovery time. The minimum open time could be less than 20% for some applications but never less than 1.0 seconds. The absolute minimum exists because of the time required for the injection valves to open, for the SCF blowing agent to begin to blow, and for the SCF flow to stabilize. The maximum injector-open time is important so that the entire injected SCF blowing agent will be able to properly mix and dissolve into the molten polymer prior to screw idling or injection.

Table 5.1 *Typical Nitrogen Levels in Various Resins*

Resin type	Typical N_2 levels (%)
Polyolefins	
HDPE	1.0 to 2.0
PP – unfilled	1.0 to 2.0
PP – talc filled	0.5 to 0.75
PP – glass fiber reinforced	0.5
Amorphous	
PS and PC	0.4 to 0.6
HIPS, ABS, IM PC	0.6 to 0.8
Glass fiber reinforced amorphous	0.3 to 0.5
Semi-crystalline	
POM – unfilled	0.2 to 0.4
POM – glass fiber reinforced	0.1 to 0.3
PA – unfilled	0.5 to 0.7
PA, PBT, PET – glass fiber reinforced	0.2 to 0.4
High-heat materials	
Polysulfone, PEEK, etc. – unfilled	0.5 to 0.7
Polysulfone, PEEK, etc. – glass fiber reinforced	0.3 to 0.4

For two injectors, the second injector should open at 2.00D (D = screw diameter) and the first injector should close sooner than 2.25D.

5.5.2 Surface Appearance Optimization

As with any foam molding process, the microcellular process cannot be expected to yield Class A surface molded parts. However, many studies have been performed to determine how to obtain the best surface appearance possible with the MuCell microcellular process.

SCF levels and injection velocities are the most important factors in optimizing molded part surface appearance. Melt and mold temperatures also have an effect, but their impact depends on the material. Also, nitrogen is generally recommended over carbon dioxide because it has a much lower rate of diffusion through the polymer, making it less likely to diffuse through the material to the flow front and to form open cells and splay on the mold surface.

5.5.2.1 Filled Semi-Crystalline Engineering Resins

Filled semi-crystalline engineering resins yield microcellular parts with an appearance almost like solid parts. To optimize surface appearance, the SCF level should be less than 0.10% and the weight reduction should be limited to 15%. Glass fiber or mineral particles facilitate cell nucleation and yield a good-quality cell structure at the lower SCF level.

The most critical parameter in surface appearance is injection speed. There is always a slight delay between the opening of the valve gates (or shut-off nozzle) and the screw moving forward during injection. During this delay, a slight "pre-foaming" occurs. When the injection velocity is high, this light colored material is randomly "sprayed" on the surface of the mold cavity.

Profiling injection velocity (as shown in Fig. 5.7) keeps this material around the gate. A typical first injection profile step might be 0.25 to 0.1.25 cm/s for 0.75 to 0.1.25 cm of stroke. With a cold runner mold, or a mold with valve gates leading into cold runners, the profile should be kept slow until the polymer enters the mold cavity. Once the flow front is established, the injection speed can be ramped up to allow for better cell structure control and cycle time. The upper end of the injection profile will probably be less than 7.5 cm/s. Some optimization is needed to determine the maximum injection speed and ramp profile. Ramping up the speed too fast may create surface wrinkling with the material cooling against the mold wall, folding in on itself and creating flow lines perpendicular to the direction of flow. Conversely, too slow a fill speed may create a visually splay-free appearance, but with a surface rough to the touch as the material may freeze before it packs out against the mold.

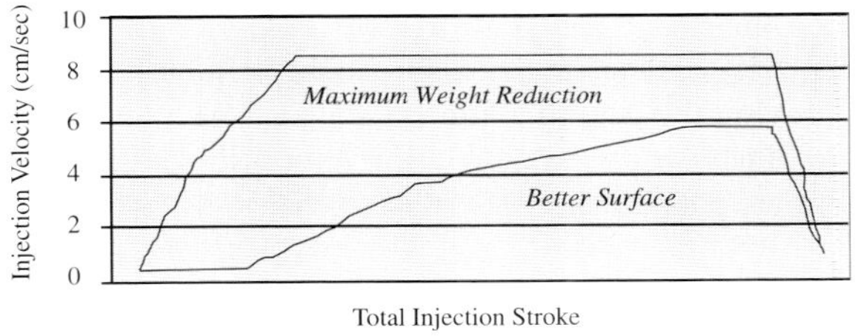

Figure 5.7 Injection velocity profiling

Temperatures also have an impact on part appearance. Higher mold and melt temperatures allow better replication of mold surfaces. Higher melt temperatures should be tried first because this will have less effect on cycle times.

Note that with unfilled semi-crystalline materials, excessively high melt temperatures may cause overly low melt viscosities and the loss of gas from the melt flow front that will negatively impact weight reduction and aesthetics.

5.5.2.2 Amorphous Resins

With microcellular processing, amorphous resins tend to take on a splayed, almost pearlescent appearance. At SCF levels as low as 0.1%, which is typical for filled semi-crystalline materials, surfaces may have a scattered splay. As the SCF level increases to 0.6 to 0.8%, the part develops a more uniform cell structure and the surface becomes uniformly lighter in color with a duller appearance.

As with filled semi-crystalline engineering resins, profiling injection velocity will affect appearance. The injection speed should start below 1.25 cm/s until the material enters the cavity. It can then be ramped up to less than 7.5 cm/s to control pre-foaming and limit splay around the gates.

With thin wall parts or small shot size parts, it may be better to fill the cavity at maximum speed with a flat profile. This will prevent the SCF from diffusing to the flow front prior to the filling stage ending.

5.5.2.3 Gating

A key to good surface finish is controlling how the SCF diffuses to the surface of the part. This makes flow length important. For optimum appearance, it is preferable to keep the flow length under 18 cm, which should be considered when selecting a gate location. Longer flow lengths are possible, but then optimizing appearance becomes more challenging. Since the microcellular process provides some latitude with the elimination of the pack and hold phases, it is often best to gate near the middle of the part to minimize the maximum flow distance.

Gates also should be located so weld line areas can vent properly. Weld lines should be avoided near the end of flow, as the surfaces in that area tend to be rougher than in the surrounding areas. Gates placed directly onto the part surface may create surface blemishes. Ideally, a hot runner will lead into a short cold runner to catch the pre-foamed material from the valve gate. The runner also should include a 'cold slug' overflow area with good venting. Moving the gate to a non-critical surface allows for a wider processing window because gate blush is not visible. Gating into a rib or ejector pin post also can be an effective way to isolate gate blush.

5.5.2.4 Additional Considerations

- Visual parts tend to achieve the best appearance when a light colorant is used. White and light gray are the colors most easily optimized for appearance. Unfilled, semi-crystalline engineering materials and amorphous polymers tend to have a lighter appearance with a splayed surface.
- A texture on the mold typically will improve the surface appearance of a microcellular part. A texture of 0.002 to 0.005 cm in depth will improve a finish, but a deeper grain may work better in some situations.
- Limiting the weight reduction to 5% or less improves the accurate reproduction of the textured surface.

5.5.3 Part Weight Optimization

When running a trial, one of the common goals is to determine the maximum weight reduction achievable for a given mold and material combination.

In order to increase weight reduction, reduce the shot size until the part is just short and then increase the SCF level until the part fills out again. Repeat this process until it is no longer possible to achieve additional weight reductions with additional increases in SCF level. It may be necessary to increase MPP and/or screw speed to maintain the molten polymer/SCF single-phase solution as SCF level approaches the solubility limit.

Once the point has been reached where higher SCF levels do not yield increased weight reduction, injection speed should be investigated. Injection velocity should be increased until there is no benefit. This will most likely result in changes to the appearance of the part.

The final variables for maximizing weight reduction are increased melt and mold temperatures. It is recommended to increase melt temperature first as this has less of an affect on cycle time. However, never run higher temperatures than the material manufacturer's recommended conditions.

5.5.4 Cycle Time Optimization

As with solid molding, cycle time in microcellular molding is controlled by the ability to remove heat from the polymer. There are two basic ways to do this: inject colder material into the mold or reduce the mold temperature.

Due to the reduced viscosity of the single-phase solution it is possible to use lower melt temperatures than with solid molding. In fact, most materials should be processable at melt temperatures as much as 20 to 40 °C below those for normal injection molding.

The materials should also be processable with a colder mold temperature. For a material, which normally runs at low mold temperatures such as PP, HDPE, and PS, it is not necessarily practical to reduce mold temperatures as this would require a chiller. However, with materials that normally run at elevated mold temperatures such as PC, PBT, and PA, mold temperatures can be reduced by 25 °C to 50 °C. Using colder mold temperatures is the single most effective way of reducing cycle time.

Finally, the SCF level should be as low as practical as this results in a lower gas pressure in the cells which also minimizes the potential for post blow.

When running thicker parts, carbon dioxide should be investigated due to its higher solubility levels.

5.5.5 Cell Structure Optimization

The best way to evaluate cell structure is to break the parts as opposed to cutting the parts. To break parts, score the surface first with a knife and break the part on the score line. Note that the cell structure at the score will be compressed, which will create the appearance of a thicker skin layer on the side that is scored. If the part is too thick to snap, break the part with a hammer. Do not cut the part with a saw as it will compress the cell structure.

There are a number of possible cell structure problems. The most common are typically cells being too large or not uniform throughout the part. Cell structure issues typically relate to either an insufficient driving force to form many small cells or to poor control of cell growth.

Insufficient driving force results in a small number of cells that do not nucleate simultaneously. Thus the few cells that are generated tend to be large and of varying sizes. The large cells can be uniformly distributed or may be in low pressure areas of the part such as the end of fill or thick sections.

Insufficient driving force is caused by either too little SCF or the pressure drop rate being too low. At higher levels of SCF, cell density will increase which results in more and smaller cells. If cell size control is critical, nitrogen will almost always be better suited than carbon dioxide due to its lower diffusion rates. The one exception being in thicker wall parts where the higher solubility levels of carbon dioxide allow for higher concentrations in the polymer which results in high heat absorption during the phase change as the SCF comes out of solution and becomes a gas.

A higher driving force or faster rate of pressure drop will have the same effect. With the injection molding process this is accomplished by faster injection speeds. An insufficient driving force could also be due to low weight reduction or slow fill.

Poor cell growth control is usually caused by either poor mold cooling so that cells continue to grow in low pressure areas or by excessive weight reduction. As with insufficient driving force, poor control of cell growth tends to leave larger cells in thick part sections or at the end of fill. This phenomenon is more common with unfilled polyolefins. Poor cell growth control leads to cell coalescence, large cells at the end of flow, and non-uniform cell structure throughout the part.

Cell structure issues can often be resolved by switching between nitrogen and carbon dioxide or by increasing injection velocity.

Finally, fillers also improve the cell structure, because a filled material always has a higher stiffness than an unfilled material which means the cell walls develop strength more quickly. Also, the filler helps to create more nucleation sites thus increasing cell density.

5.5.6 Clamp Tonnage Optimization

Clamp tonnage optimization is rarely a primary goal of a processing trial. In addition, when using the microcellular molding process, clamp tonnage is normally reduced well beyond the ability of the mold to fit in a machine of the appropriate clamp size.

However when clamp tonnage reductions are critical, the pressure required to fill the mold should be reduced. On injection, this typically means profiling injection velocity to eliminate hydraulic pressure spikes at the start of injection and just before transferring out of injection. Hold pressure, if used, should be as low as possible but no lower than the MPP. Also, weight reductions should be about 10%; normally, weight reduction in excess of 10% does not result in further clamp tonnage reduction.

In terms of part geometry, thin areas at the end of fill, large flow imbalances, and poor venting should be avoided as these conditions will all result in a need for higher injection and hold pressure to fill the part which in turn will affect clamp tonnage.

5.5.7 Minimization Part Warpage

Microcellular molding can be used to minimize warpage, which is normally caused by differential shrinkage. The most common causes of the warpage are residual stress in the part, differential cooling due to mold temperature differentials, and part geometry.

The MuCell microcellular molding process reduces residual stress through elimination of hold time and pressure and by uniformly applying packing pressure through the expanding cells. As little as 2 to 5% weight reduction will reduce the residual stress enough to dramatically reduce the part warpage.

Differential cooling issues in the mold must be taken care of through mold and/or part design.

5.5.8 Summary

Table 5.2 summarizes the actions for optimizing the MuCell microcellular process. The actions are listed in order of highest to lowest priority.

Table 5.2 Summary of Microcellular Molding Optimization Actions

Goal	Action
Minimize part weight	Reduce shot size and then increase SCF level Increase MPP if necessary Increase screw speed if necessary Repeat first three steps until no further reduction obtained Increase injection speed Increase mold and melt temperatures
Minimize cycle time	Decrease mold and melt temperatures Decrease SCF level Decrease screw speed Switch to carbon dioxide for thick parts
Optimize cell structure	Use nitrogen Increase SCF level Decrease weight reduction Increase injection speed Minimize mold and melt temperatures Add 5 to 20% filler
Minimize clamp tonnage	Maximize weight reduction Minimize pack and hold pressure Decrease injection speed Minimize peak cavity pressure
Optimize surface appearance	Use nitrogen Decrease SCF level Profile injection speed slow to fast Increase melt and mold temperatures Add filler to material Texture mold
Minimize part warpage	Eliminate pack and hold time Minimize weight reduction

5.6 Troubleshooting the Microcellular Molding Process

Other issues associated with MuCell parts and processing can be broken down into two categories: surface appearance issues and process issues.

5.6.1 Surface Appearance Issues

As stated previously, the MuCell microcellular molding process cannot be used to produce a Class A surface. Earlier it was discussed how to produce a uniform surface appearance. This section will address other potential appearance problems. These include blistering, both internal to the part or on the surface as well as blowouts, dimples, and large surface defects.

5.6.1.1 Internal Blisters

As seen in Fig. 5.8, internal blisters have the appearance of a bubble on the part and in an extreme case the blister may burst. Typically internal blisters are 6 to 10 mm in diameter; sometimes they are larger.

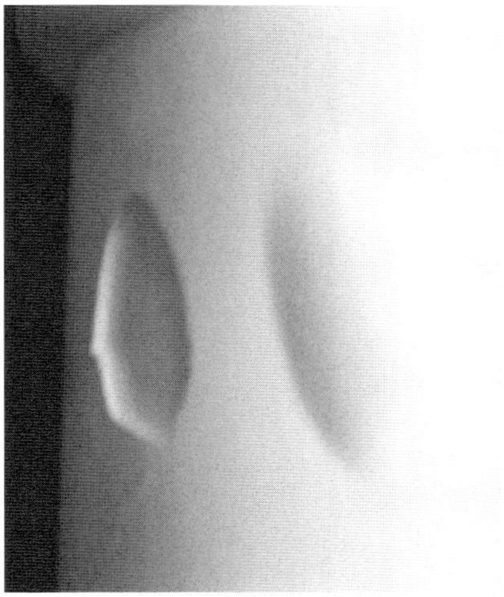

Figure 5.8 *Internal blisters*

In order to properly diagnose an internal blister, the bubble should be cut open. If the bubble consists of a gas pocket with a smooth wall surface and a relatively thick skin, this is an internal blister. It is caused by the SCF being out of solution such that a void is injected into the part. The excessive pressure in this void caused the walls to expand. Increasing cooling time will not eliminate the bubble.

This issue is common to unfilled crystalline materials such as polyolefins and acetal but can also occur in ABS.

The primary cause for internal blisters is that SCF is not in solution in the plasticator barrel. This can be corrected in a number of ways. Decreasing the SCF level is the best approach, however, increasing the MPP or the melt temperature also helps.

The other possible cause for internal blisters is that the SCF level is low and the cell structure is large. In this case, the SCF level needs to be increased. This cause for internal blisters is prevalent in HDPE and PP.

5.6.1.2 Surface Blistering

Surface blistering can best be described as small (1 mm or less) bubbles on the surface of the part. These bubbles are different from internal blisters in that these bubbles have a very thin skin and are normally seen near the gates. Surface blistering is more common with unfilled polyolefins but has been seen with PS, acetal (POM), and unfilled polycarbonate.

The most common cause of surface blistering, especially in amorphous resins, is high shear going through the gates. Slowing injection speed at least until the material enters the mold cavity can eliminate this problem.

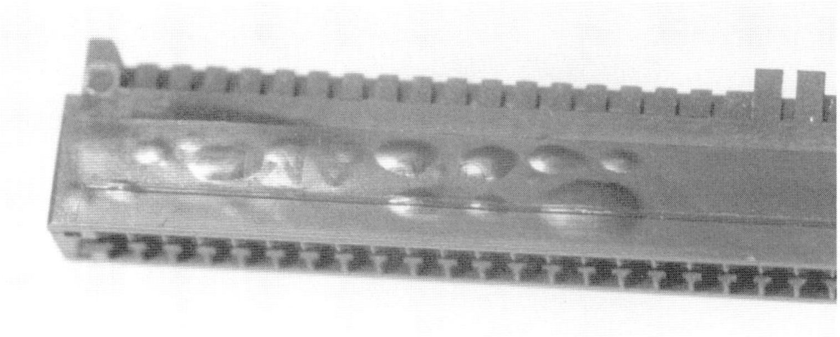

Figure 5.9 *Surface blisters*

Another cause for surface blisters that is less common is insolubility of the blowing agent in the molten polymer. In this instance, a switch to carbon dioxide should be evaluated if nitrogen is being used then.

5.6.1.3 Blowouts

Blowouts or post blow (see Fig. 5.10) appear to be the same as an internal blister from the outside. However, if the bubble is cut open, the inside will have a cell structure either elongated or damaged as shown in Fig. 5.11.

A good way to differentiate between a blowout and an internal blister is to extend cooling time. Blowouts will be reduced or eliminated by longer cooling times where as an internal blister will not be affected.

Figure 5.10 Blowout

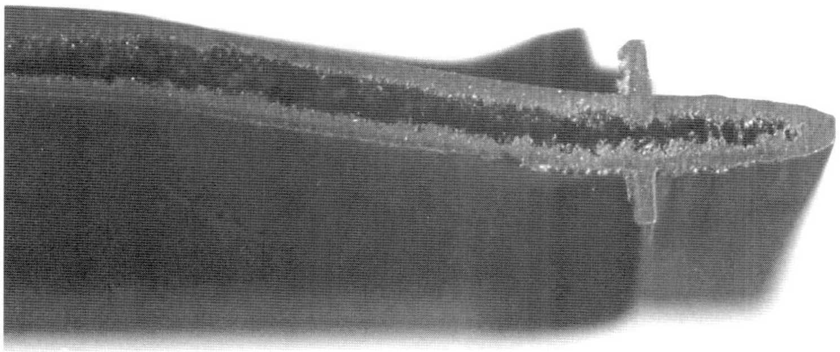

Figure 5.11 Cutaway of Blowout

The most common cause of blowouts is a hot spot in the part. Blowouts occur because the material is not rigid enough in that particular location to withstand the residual gas pressure in the cells when the mold opens. Blowouts can be corrected by decreasing mold temperature, extending cooling time, or reducing SCF level. However, it should be remembered that reducing the SCF level could create problems with cell structure.

5.6.1.4 Dimples

Dimples are small depressions in the surface that can be about 1 to 2 mm in diameter. These depressions tend to form over large cells that are near the surface. Dimpling is an indication of excessive weight reduction for the SCF level being used. Therefore, dimpling is usually resolved by increasing the SCF level or by decreasing the weight reduction.

5.6.2 Process Issues

Most process related issues are related to the SCF blowing agent coming out of solution, running close to the blowing agent solubility limits of the polymer, or tooling issues.

5.6.2.1 SCF out of Solution

When the SCF blowing agent comes out of the single-phase solution in the injection molding barrel, there are usually numerous symptoms. The most obvious symptom is random short shots occasionally interspersed with full parts. Often the short area will not be in the same location on the part; the part will be significantly short, or the short area may not be at the end of flow. Often the flow front is rough, friable, and has a white powder on it when the part is removed from the mold.

Other symptoms include

- Internal blisters
- Shot weight variation of as much as 40%
- Large variations in recovery time

Significant variations in SCF inlet pressure and in injection pressure

The best corrective action is to back the barrel away from the mold, purge it empty, and to start over. Recalculate the SCF level that was going into the barrel and start at about 50% of the previous SCF level. The corrective actions for SCF coming out of solution are the same as discussed previously for internal blisters.

It is also possible that small amounts of SCF have failed to mix into the molten polymer and shows up randomly in the molded part. Often, this is caused by a large drop in delivery pressure, which leads to a surge of SCF flowing into the plasticator barrel when the SCF injector opens. This is corrected by keeping the delivery pressure drop to less than 15 bar.

5.6.2.2 Short Shots

Any gas trap issue with the mold using the conventional solid molding process will be aggravated with the MuCell process. A poorly vented mold allows gas traps to occur when the air in front of the material flowing into the part is unable to vent out of the mold cavity. The processor then typically uses excessive pressures to fill the part. In solid molding, this leads to higher molded-in stress and cosmetic flaws. In microcellular molding, this limits the weight reduction that is possible.

There are a couple of good signs that there is a venting or gas trap problem with the mold. If most of the mold fills with good weight reduction, such as 10% or 15%, but the final areas do not fill out with these same weight reductions this is an indication of a venting problem. The other possibility is to decrease the clamp tonnage. If this allows for significantly greater weight reductions then venting is an issue.

There are a number of ways to correct short shots. Clamp tonnage can be reduced, injection speed can be reduced or profiled, the mold venting can be improved, or rib thicknesses can be increased (gas trap in a rib). Hold pressure can also be used to correct gas trap problems just as in the solid process; however this will add weight back to the part.

5.6.2.3 Inconsistent Part Weight

During standard injection molding, hold pressure and cushion are used to compensate for variations in the molding process and material. The MuCell process does not use hold pressure. Therefore, the parts may have slightly more weight variation than those produced by a solid process. Often the microcellular molding process has 2 to 3 times more variation than the conventional process. Therefore, if the solid part is running +/- 0.1 g, the MuCell part will run +/- 0.2 to 0.3 g. In solid molding, the weight variation is often used to indirectly monitor dimensional variation. This is because both the final part weight and the part dimensions are controlled by the pack and hold phases of the mold filling process. The final mold shrinkage is closely related to the pack and hold pressures.

With microcellular molding, the final part dimension relies on cell growth and thus the shrinkage is not changed by the packing pressures across the mold cavities. On a microscopic level, the average cell size changes slightly as the weight changes but the shrinkage remains more constant than in solid molding.

To minimize weight variation in the MuCell process, the shot size and the transfer point must be consistent. This can be achieved by profiling screw speed to slow just prior to reaching the shot size and also profiling injection to slow just prior to the transfer point. This will help provide more consistency in the starting point and the ending point of the injection stroke.

If the weight inconsistency still persists, the SCF level, MPP settings, and screw speed should be evaluated. Normally, the SCF level should be lowered. The MPP should be increased if the screw is drifting backward after screw recovery or decreased if the screw is drifting forward.

5.6.2.4 Inconsistent Screw Recovery

Screw stalling and inconsistent screw recovery are caused by similar issues. The cause is either SCF coming out of solution in the barrel or a very high level of SCF with high MPP which makes it difficult for the screw to recover, particularly with low-viscosity materials.

If the current SCF level and MPP must be maintained, try increasing screw speed. However, if the problem persists, back the screw off and purge out as there is most likely SCF out of solution in the barrel.

References

1. Kwag, C., Manke, C.W., Gulari, E., *J. Polym. Sci.: Part B: Polymer Physics* (1999), pp. 2771–2781.

6 Microcellular Molding: Molded Part Properties

6.1 Background

As expected, microcellular molding affects physical properties of molded ASTM or ISO test specimens. The addition of small foam cells within the solid polymer matrix decreases measured physical properties. With chemical blowing agents in standard foam molding, the loss in physical properties can occur geometrically with density reduction. On the other hand with microcellular molding, the loss in properties is usually more linear with density reduction.

In actual microcellular molded articles, physical properties may actually improve due to lower stresses, more uniform packing of the part due to foam expansion, and high skin to core thickness ratios due to lower melt temperatures, and cooler molds. Much work has been done to understand the effect of microcellular molding on properties. Many of the results are presented in the following together with a model for predicting some of the physical properties.

6.2 Effect on Surface Appearance

The surface appearance of a part molded with the MuCell process will demonstrate splay. The splay seen in MuCell parts is most likely caused by the gaseous blowing agent exiting the melt front as it impinges on the cavity wall; a small amount of the gas may then occasionally get trapped between the mold surface and the material. In addition, small cells may be formed at the interface between the melt and the cavity wall, which are then dragged, stretched, and opened until the melt cools. However, the level to which this occurs and the uniformity of the surface appearance can be modified through processing conditions. It should be noted that even under the best circumstances, the MuCell molding process is not recommended for producing parts with a high-quality Class A surface. Considering this, the process is best suited for non-appearance parts, either structural or as the rigid material behind a decorative fabric or film.

The microcellular molding process often does improve some surface defects including sink marks. Sink marks are usually reduced, if not eliminated, because the microcellular molding process packs out the part due to the foaming process in the mold.

6.3 Effect on Shrinkage

The shrinkage properties of all materials but particularly of semi-crystalline materials are highly dependent on wall thickness and packing pressure. The shrinkage in the part will vary if packing pressure varies. The advantage of the MuCell process is that the pressure of the gas in the cells is the packing pressure and this is essentially uniform throughout the part. Because the packing pressure is uniform, the shrinkage is also more uniform, resulting in a significant reduction in part warpage or bowing. Thus, the MuCell process is useful for parts requiring superior dimensional compliance, especially when multiple molded parts need to be assembled.

Figure 6.1 presents shrinkage values from ISO tensile bar tests for 4 different grades of PP: an unfilled copolymer, a 20% talc-filled copolymer, a 20% glass fiber-reinforced copolymer, and a 40% glass fiber-filled copolymer. The shrinkage values for the glass fiber reinforced materials and the unfilled material did not show a change. There was however, a slight change in the talc-filled material as the part weight was reduced although the difference of 0.002 mm/mm (0.2%) of length was within the normal window provided by the material suppliers.

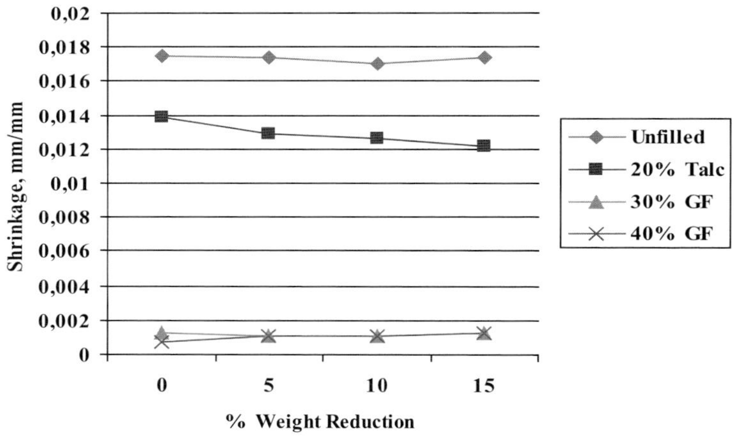

Figure 6.1 *Comparison of shrinkage (PP)*

Figure 6.2 presents shrinkage values from ISO tensile bar tests for four different impact modified amorphous materials. As seen in the figure, the shrinkage values for PPO/HIPS did not show a change. However, for the PC-based materials the shrinkage value changed as soon as the weight was reduced. For ABS, the shrinkage gradually increased as the level of weight reduction increased. It appears that the shrinkage of single-phase materials (PPO/HIPS) is less affected by the MuCell process than those of multi-phase materials. The cause of this phenomenon has not yet been identified .

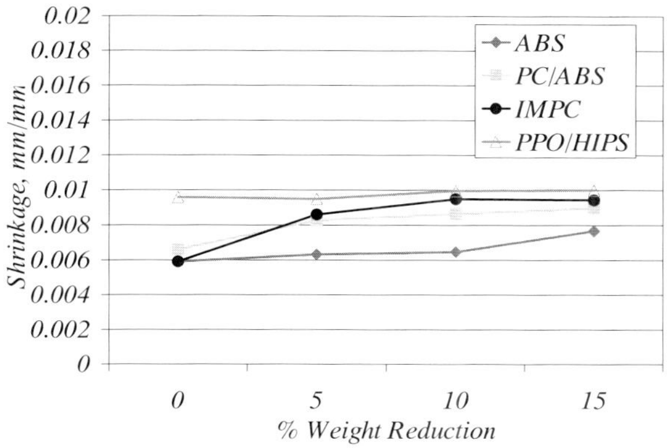

Figure 6.2 *Comparison of shrinkage (impact modified amorphous materials)*

An additional advantage of the microcellular molding process is that the dimensional stability of molded parts in thermal cycling is improved due to the reduced molded-in stress.

6.4 Flexural, Tensile, and Izod Properties for Polypropylene

Figures 6.3 through 6.7 show ISO property data for four different polypropylene materials: an unfilled impact copolymer, a 20% talc-filled copolymer, a 20% glass fiber-filled copolymer, and a 40% glass fiber-filled copolymer. Each was evaluated using standard ISO bar test specimens; the test samples were molded based on ISO 294:1995 and were tested based on ISO 3167.

Under these conditions, it can be seen that flexural properties are relatively unaffected by changes in weight reduction. This relative insensitivity of the flexural modulus to weight reduction is a primary reason why the process is effective for structural applications. The other benefit in structural parts is the fact that relatively small changes in flexural modulus have little affect on overall part performance.

Tensile properties show a different behavior. Typically, there is an initial drop at 5% weight reduction, then a leveling-off of material performance. At 20% or more, this initial drop in tensile strength is quite significant. This is not observed in the flexural properties because those are much less dependent on cell structure and more dependent on the skin to core ratio, which tends to be unaffected by weight reduction.

Notched Izod impact performance shows the same trend as tensile properties but the initial drop with the unfilled material is larger. This 40% drop in impact strength for unfilled polypropylene is relatively common for ductile materials. In filled materials, the characteristics of the filler tend to control the impact performance of the product, and the MuCell process does not have an affect on the fillers. Because of this, and the fact that the materials tend to be more brittle, impact property retention is better.

Given the nature of these changes in tensile and impact properties, it is recommended for most applications that weight reductions of no more than 10% to 15% be used when running with polypropylene materials.

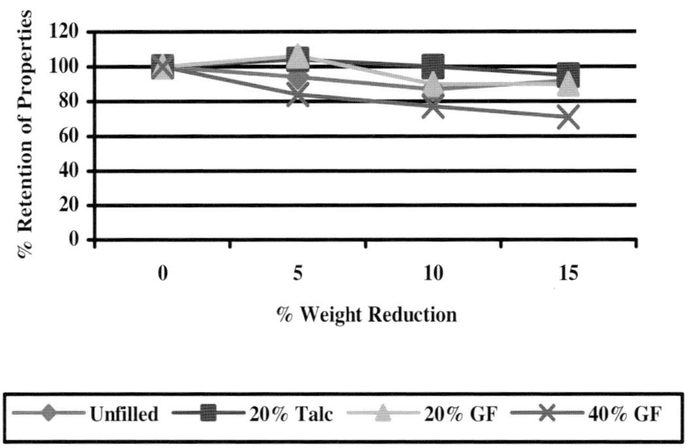

Figure 6.3 *Flexural modulus*

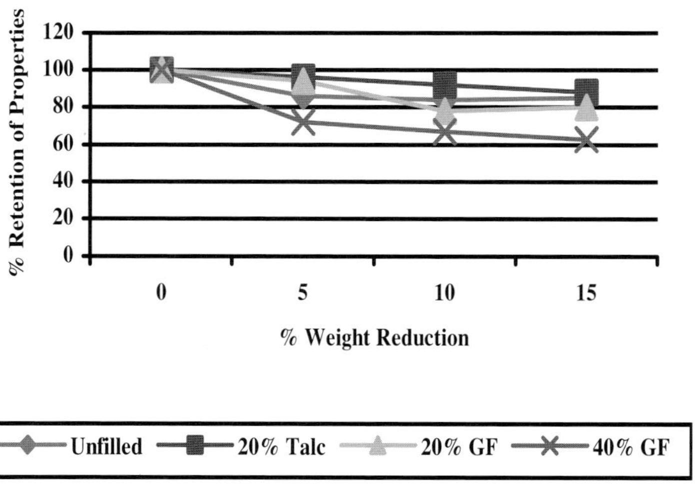

Figure 6.4 *Flexural strength*

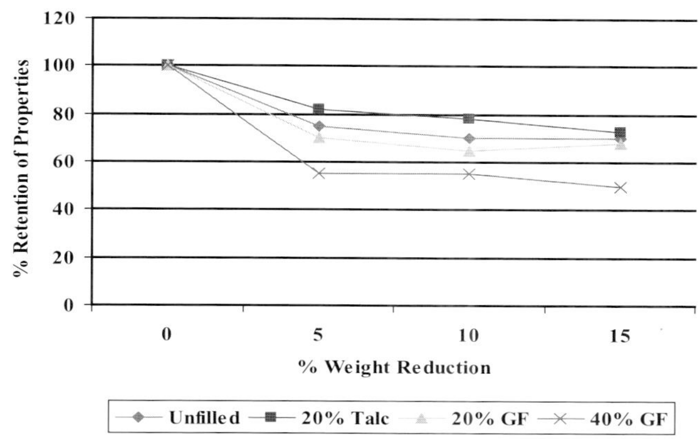

Figure 6.5 *Tensile strength*

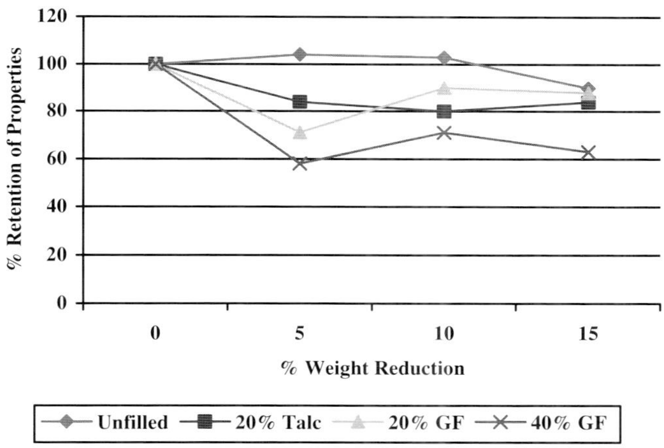

Figure 6.6 *Tensile elongation at yield*

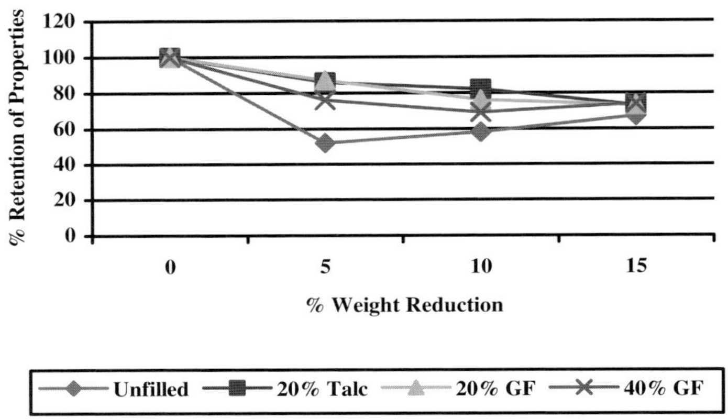

Figure 6.7 *Notched Izod impact strength*

6.5 Physical Property Changes by Material

Numerous materials other than polypropylene have been tested for the effect of microcellular molding on physical properties. In the following figures, the properties of selected MuCell and conventionally molded parts are compared and presented in radar charts. The level of weight reduction of the parts processed with the MuCell process was varied from 2 to 20%.

In Figures 6.8 through 6.13, the test results presented are scaled relative to conventionally molded specimens and are therefore dimensionless. The dimensionless properties were calculated as follows:

$$P_{dimensionless} = \frac{P_{MuCell} / m_{MuCell}}{P_{Conventional} / m_{Conventional}} \tag{6.1}$$

where $P_{dimensionless}$ is a dimensionless part property, P_{MuCell} is a MuCell processed part property, $P_{Conventional}$ is a conventionally molded part property, m_{MuCell} is a MuCell processed part weight, and $m_{Conventional}$ is a conventionally molded part weight.

The scale in the charts indicates whether the part property is changing proportional to weight reduction. If the scaling number is less than 1, the part property is reduced by some value more than the weight reduction. Similarly, if the scaling number is greater than 1, the part property decreased less than the percent weight reduction.

The discussion of the test results is divided into two categories by material types: ductile/tough (as seen in Figures 6.8 through 6.10) and brittle/fiber-filled (as seen in Figures 6.11 though 6.13).

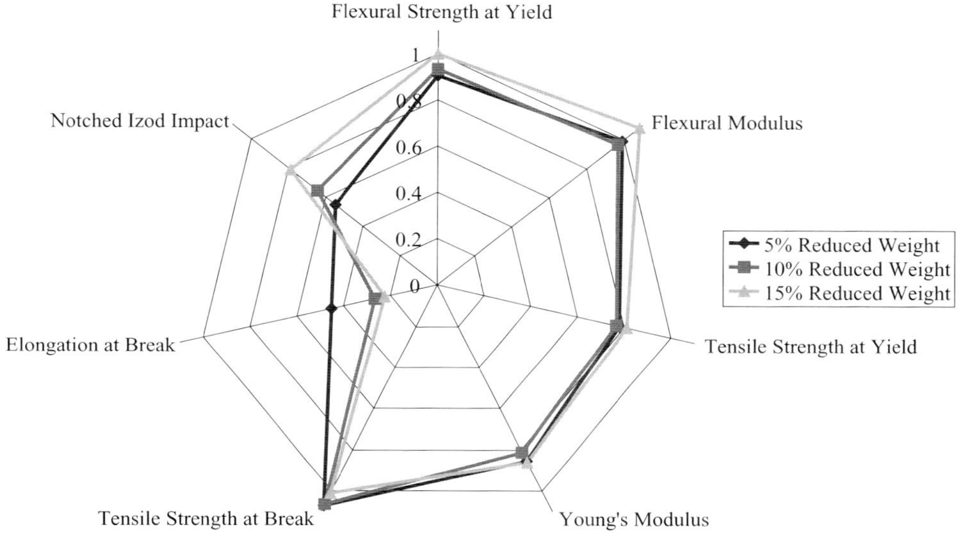

Figure 6.8 *Property comparison DOW Inspire MFR12 PP*

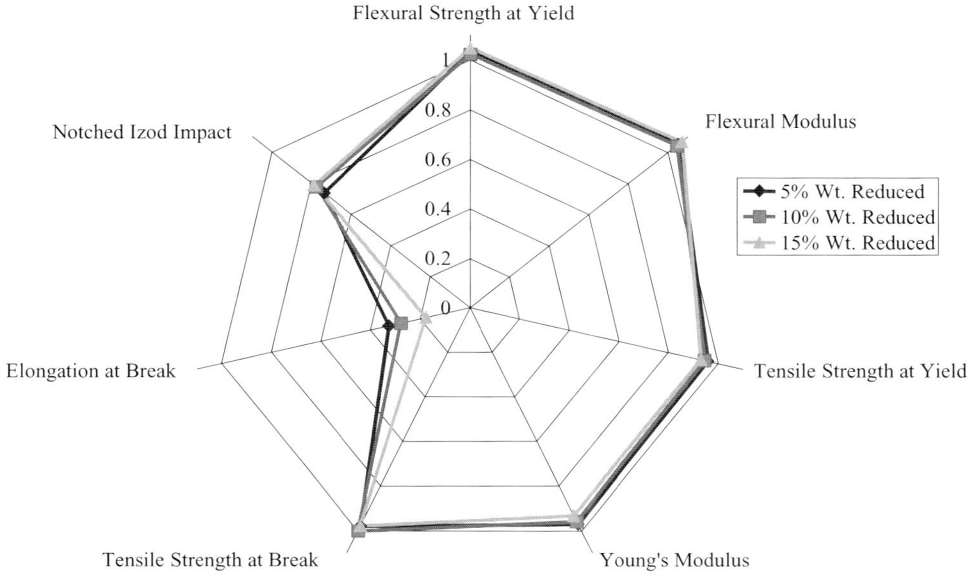

Figure 6.9 *Property comparison GE Lexan IP300 PC*

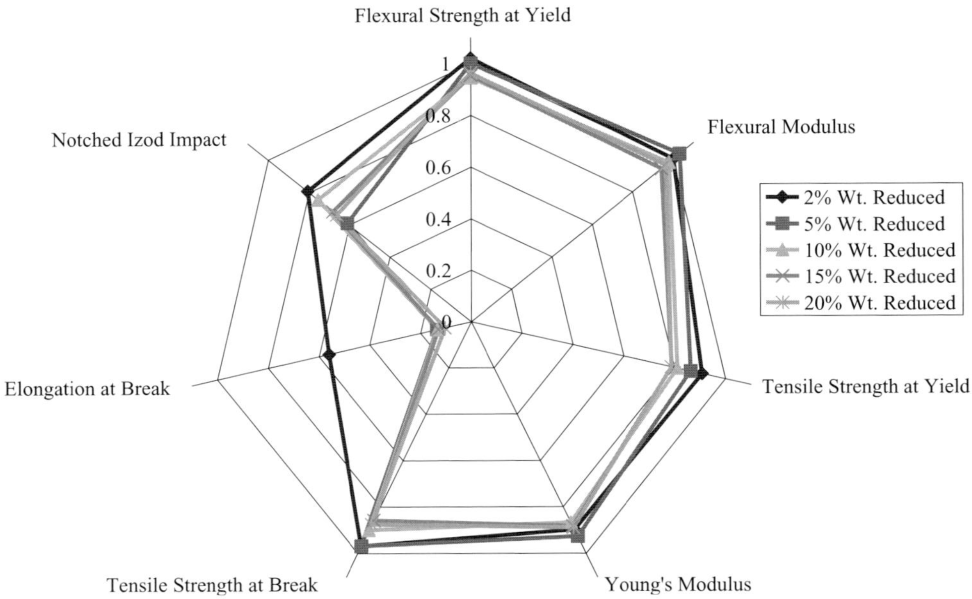

Figure 6.10 *Property comparison GE Noryl 731 SEI PPO/PS*

For Figures 6.8 through 6.10, it is clearly seen that the flexural properties per unit weight, except for PP-based parts, are equivalent to or even higher than those of conventionally molded solid parts. The tensile strength at yield and modulus per unit weight of MuCell parts drops more rapidly than the flexural properties but is still in the normalized range of 0.8 to 0.9, again except for PP-based parts. However, the toughness of these ductile materials was significantly reduced as compared with the conventionally molded parts.

A material's mechanical properties decrease as density decreases and under the best conditions the properties will give a ratio of 1 when normalized for density. However, MuCell parts do not fully follow this. For example, a significant reduction in toughness was observed with MuCell parts, which was not only caused by weight/density reduction but also due to other causes. On the other hand, the relative strength did not show much reduction and in some cases, flexural properties showed slight improvements as shown in Figures 6.8 through 6.10.

The tensile and flexural strength strongly depend on the skin thickness, as well as on cell sizes. The higher skin thickness and smaller cell sizes result in less strength reduction. This can be explained by the concept of beam bending. A molded part with a thick skin can be thought of as a hollow beam with a thick wall. On the other hand, a molded part with a thin skin can be thought of as a hollow beam with a thin wall. Detailed explanation for the tensile and flexural strengths can be found in the next section.

Cell structure dictates the properties of the unfilled microcellular molded materials. It appears that the cell formation and surface finish might have affected the part properties. This is very critical, particularly for polycarbonate-based parts because PC is very sensitive to notches or scratches. The cells and the surface roughness in the MuCell parts probably acted as stress concentrators and created an inherent discontinuity in the parts, reducing toughness.

For Figures 6.11 through 6.13, the trend in part properties of brittle materials was different from those of ductile materials. As with the ductile materials, tensile and flexural properties per unit weight of fiber-filled MuCell parts were almost equivalent to those of conventionally molded parts. However, unlike the ductile materials, the toughness of these MuCell processed fiber-filled materials did not show much reduction as compared with the ductile materials. In the case of glass fiber-reinforced PBT, toughness was actually improved.

This can be attributed to the fact that the performance of fiber-filled materials mainly depends on the type, amount, and orientation of reinforcements, as opposed to the polymer matrix. In the case of glass fiber-reinforced polymers, the glass fiber and the adhesion between the glass fibers and the polymer control the property performance. As this is not significantly affected by the cell structure, there is less property change.

The other benefit of glass fiber-reinforced materials is that they are brittle. The most significant affect of cell structure on ductile materials is in toughness. Brittle materials do not have the elongation properties and therefore, this is not an issue.

It should be emphasized that the MuCell part properties shown here are based on the ISO bars (thickness 4 mm), which made it more difficult to maintain smaller cell sizes. Therefore, actual molded parts with thinner thicknesses are expected to retain more of the molded solid part properties than described in the preceding.

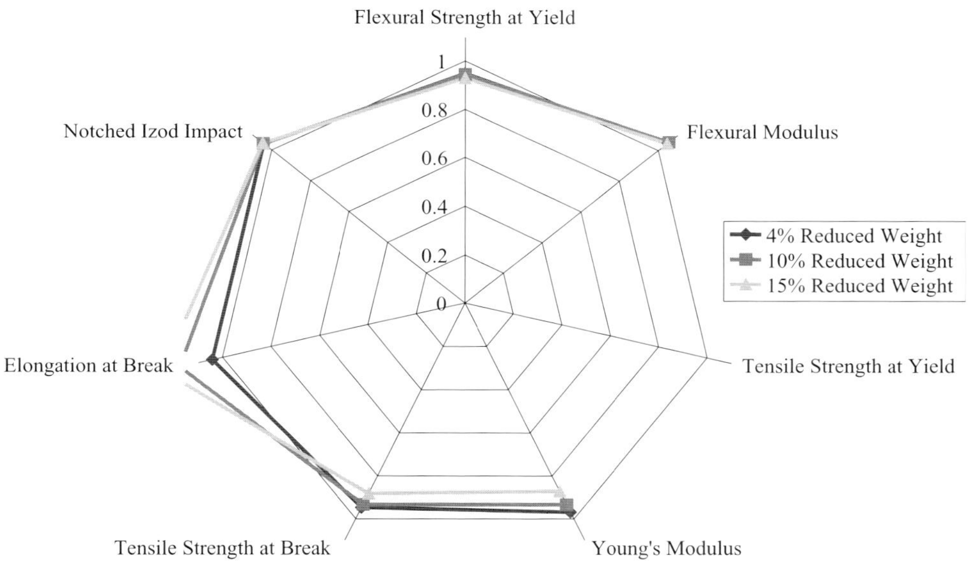

Figure 6.11 *Property comparison GE Valox 420 SEO 33% GF-PBT*

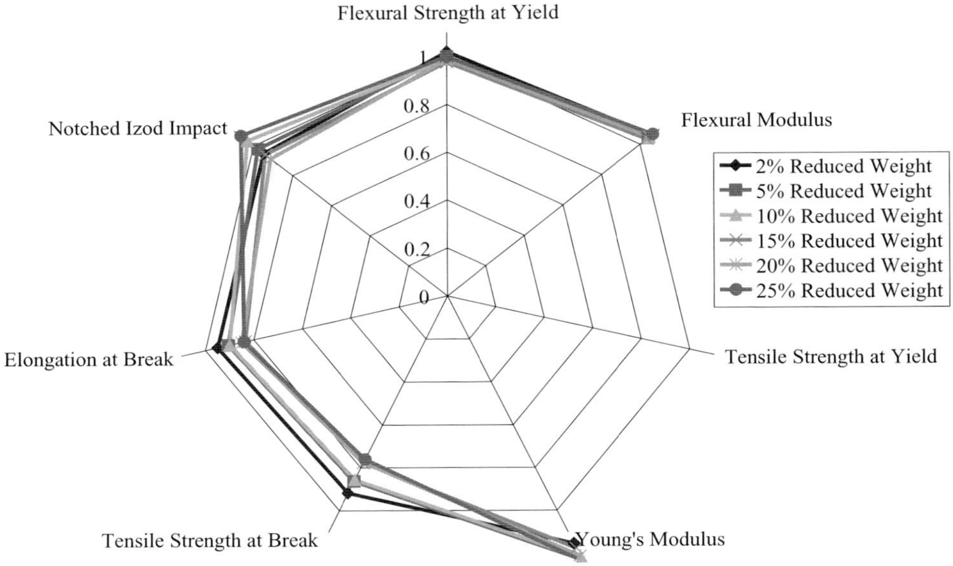

Figure 6.12 *Property comparison DuPont Zytel 73G33HSLBKB031 33% GF-PA*

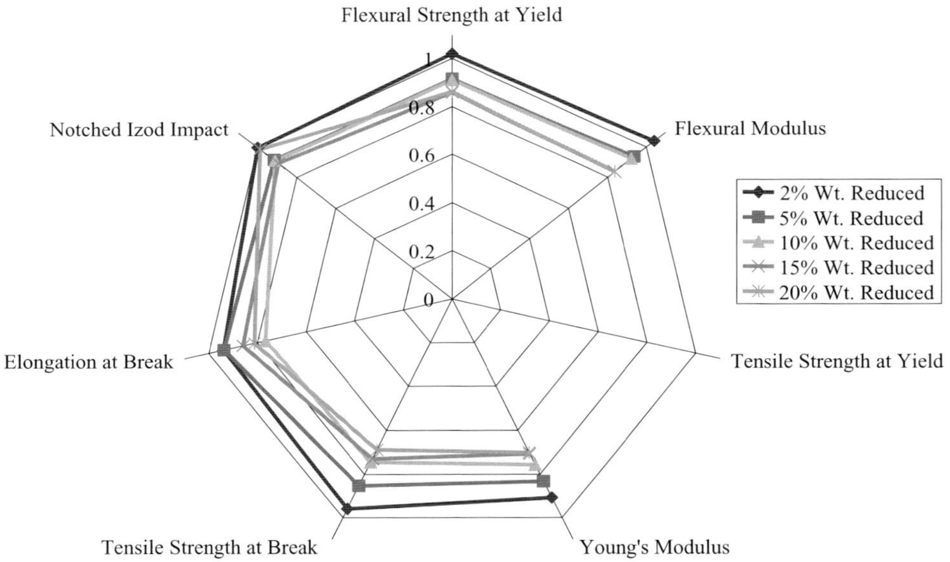

Figure 6.13 *Property comparison GE Noryl GFN2 20% GF PPO/PS*

6.6 Physical Property Modeling

Many efforts have been made to predict mechanical properties based on microcellular cell morphology using a complicated mathematical model [1-3]. Since microcellular molding can control weight reduction, a sandwich structure model was developed that describes a microcellular molded article as a uniformly foamed core encased by a skin frame. Figure 6.14 shows a typical SEM of a cross-section view for an unfilled PC microcellular molded article. The average cell size of the sample in Fig. 6.14 is about 45 microns.

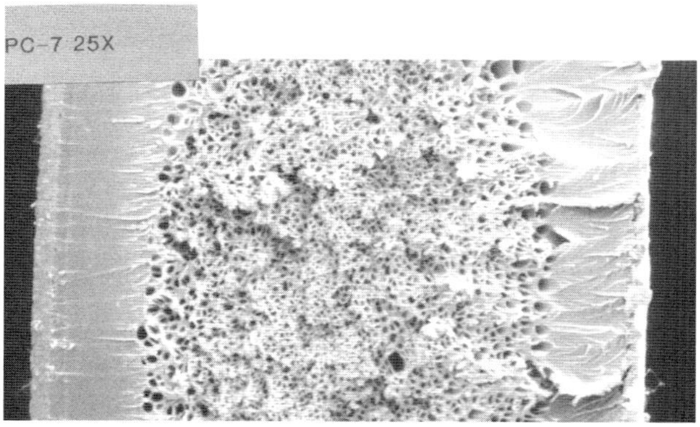

Figure 6.14 *SEM of unfilled PC sample; thickness: 3.7 mm; skin thickness: 0.65 mm; mold temperature: 160 °F; melt temperature: 580 °F; pressure drop rate: 1.7×10^{11} Pa/s*

Microcellular injection molded parts have the unique characteristic of a uniform density profile across the foamed core. It is different from the bell-shaped density profile that exists across the thickness of a conventionally foamed part, which typically has large cells in the center and small cells near the skin [4].

The cell morphology effect for the tensile and flexural strength calculations is negligible and accounts for less than 10% error in property predictions if the cell size is equal to or less than 100 microns and the cells are distributed uniformly. Therefore, an easy prediction model was developed that using only weight reduction percentage and skin thickness as the input data to calculate the final mechanical properties of tensile and flexural strength for microcellular molded parts. However, impact strength modeling requires that cell morphology be taken into account.

6.6.1 Model for Tensile Strength

The models for unfilled materials and filled materials are different because of the relationships between tensile strength and weight reductions.

6.6.1.1 Unfilled Materials

To simplify the mathematical model, the following assumptions are made:

1. A sandwich model is proposed for a microcellular part, i.e., a uniform foamed core is encased by a uniform solid skin.
2. The average cell size is equal to or less than 100 microns.
3. In tension, the entire cross section of the sample is subjected to a uniform stress.

With these assumptions, a tensile strength ratio R_{tu} is defined as:

$$R_{tu} = \frac{bh - (h-2t)(b-2t) + (1-R)^2(h-2t)(b-2t)}{bh} \tag{6.2}$$

Where

R_{tu} is the ratio of tensile strength of unfilled foam to unfilled solid.

b is the width of the sample.

h is the whole thickness of the sample.

t is the average skin thickness of the sample.

R is the ratio of weight reduction of foamed core.

A ratio of R_w is defined as the weight reduction ratio of the whole microcellular part. Usually, R_w is known from the molded microcellular part. Then, R can be calculated:

$$R = \frac{bh(1-R_w)}{(h-2t)(b-2t)} \tag{6.3}$$

The predicted data in Table 6.1 matches the measured data well for the unfilled PC samples at 8% and 13% weight reductions with the error percentages only 1.2% and 2.6%, respectively. The average skin thickness is about 0.02 in. and the average cell size is 45 to 100 microns.

For tensile strength loss, the skin thickness may not be a very important factor if a certain total part weight reduction percentage must be maintained. For example, if the total part weight reduction is maintained at 10% but the skin thickness and core weight reduction are changed accordingly then, even if the skin thickness increases about four times, the tensile strength of the foamed part with thick skin increases only 6% while the foamed core weight reduction must be changed from 11.7% to 28.3%.

On the other hand, if the skin thickness increases about four fold with the same 10% weight reduction of the core, then the tensile strength loss of the total molded part decreases 10.8%. However, the total weight reduction in this case is also decreased to about half of the original weight reduction. In addition, the larger number of smaller cells will further increase the measured tensile strength. Therefore, the economical way to increase the tensile strength without compromising weight reduction and cycle time is to make the cells smaller and the cell density higher.

Table 6.1 *Tensile Strength Ratio Comparisons: Unfilled PC*

	Sample	
	1	**2**
Weight reduction, %	8	13
Measured R_f	0.850	0.770
Predicted R_f	0.840	0.750
% Error of prediction	–1.2	–2.6

6.6.1.2 Filled Materials

For these calculations, fillers are defined as both fillers (such as talc and calcium carbonate) and glass fiber that will not foam and thus change specific gravity. Therefore, when we calculate the weight reduction of the core material, the filler weight must be deducted from the real weight reduction calculation of the foamed core. In addition to the assumptions for unfilled resins, some extra assumptions for calculations for filled resins are:

- Filled materials are uniformly distributed in the foamed core.
- After foaming the weight percentage of filled resin in the core decreases with the same ratio *R*.

Then, the real weight reduction ratio R_{gf} of the core material without filler will be:

$$R_{gf} = \frac{\rho_s - (1-R)\rho_{sf}(1-W_{gf}R)}{\rho_s} \tag{6.4}$$

Where

R_{gf} is the real weight reduction ratio of the core material without filler.

W_{gf} is the weight percentage of filler before foaming.

ρ_{sf} is the density of filled solid material.

R is the real weight reduction ratio of the core material with filler and it can be calculated by Eq. 6.3.

A similar equation can be derived for the strength ratio of a foamed part with filler to the non-foamed part with filler:

$$R_{tf} = \frac{bh - (h-2t)(b-2t) + (1-R_{gf})^m (h-2t)(b-2t)}{bh} \tag{6.5}$$

Where

m is different for different filled materials:

for glass fiber, m equals 1.2 and

for fillers m equals 2.

The input data required for the above calculation is only the whole part weight reduction R_w, the skin thickness t, the width of the sample b, and the whole thickness of the sample h.

The comparison of the predicted versus the actual properties for filled polypropylene is shown in Table 6.2. The results of predicted tensile strength loss versus weight reduction are acceptable with percentage error of prediction between 3.4% and 6.0%. The PP sample had an average cell size of less than 100 microns and an average skin thickness of 0.015 in.

Table 6.2 *Tensile Strength Ratio Comparisons: 20% Talc-Filled PP*

	Sample		
	1	**2**	**3**
Weight reduction, %	5	10	15
Measured R_f	0.820	0.770	0.710
Predicted R_f	0.869	0.801	0.734
% Error of prediction	5.98	4.02	3.38

The prediction error percentage of fiber-filled PBT is less than 6.9% (see Table 6.2). The PBT samples with 30% glass fibers have a uniform cell structure and small cell sizes (10–50 microns) and an average skin thickness of 0.01 in. The cell growth may help to improve the glass fiber disorientation during mold filling and to increase the mechanical properties of microcellular foamed glass fiber-reinforced parts.

Table 6.3 *Tensile Strength Ratio Comparisons: 30% Glass Fiber Filled PBT*

	Sample		
	1	**2**	**3**
Weight reduction, %	4	10.910	15
Measured R_f	0.910	0.840	0.750
Predicted R_f	0.879	0.838	0.802
% Error of prediction	–3.41	–0.24	6.93

Figure 6.15 shows the results of the prediction of tensile strength loss versus whole part weight reduction percentage with the tensile strength model above. The total skin thickness in Fig. 6.15 is about 12% of whole part thickness for both unfilled and filled materials. The results in Fig. 6.15 show the tensile strength loss of the glass fiber reinforced material will be less than the one of unfilled material at the same whole part weight reduction. Generally, the weight reduction percentage should be controlled to no more than 10% for unfilled materials and to no more than 20% for glass fiber-reinforced materials if less than 20% tensile strength loss is allowed.

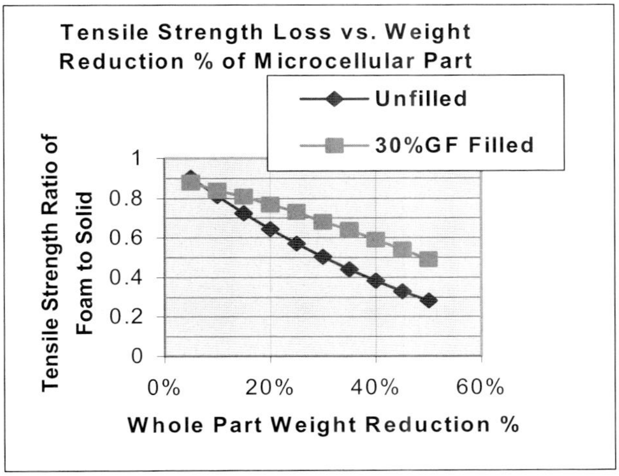

Figure 6.15 *Tensile strength loss versus whole part weight reduction % for filled and unfilled materials*

6.6.2 Model for Flexural Strength

The model for flexural strength is similar to the model for tensile strength. Once again, a different formula is used for unfilled and filled materials. The effect of skin on the side (vertical effect) is not important for flexural strength and is thus neglected. Therefore, only the thicknesses in the thin cross-section are considered.

6.6.2.1 Unfilled Materials

A flexural strength ratio R_{fu} is defined as:

$$R_{fu} = \frac{2t^3}{h^3} + \frac{6t(h-t)^2}{h^3} + (1-R)^2\left(\frac{h-2t}{h}\right)^3 \tag{6.6}$$

Then, the real weight reduction in the foamed core is R and can be calculated with Eq. 6.7:

$$R = \frac{h(1-R_w)}{h-2t} \tag{6.7}$$

The predicted data in Table 6.4 agree well with the experimental data of unfilled PC samples at 8% and 13% weight reductions with the prediction error percentage only 4.1% and 0.6%, respectively.

Table 6.4 *Flexural Strength Ratio Comparisons: Unfilled PC*

	Sample	
	1	**2**
Weight reduction, %	8	13
Measured R_f	0.980	0.910
Predicted R_f	0.940	0.905
% Error of prediction	– 4.1	– 0.6

For flexural strength, skin thickness is a very important factor if the weight reduction must be maintained for the total foamed part. With the same example as used for tensile strength, if the skin thickness increases about four times, then the flexural strength loss of the molded part decreases significantly by about 11.5%, which is double the rate of tensile strength with the lower density foamed core but same whole weight reduction.

6.6.2.2 Filled Materials

The additional assumptions here are the same as for the tensile strength calculation. A similar equation can be derived for the strength ratio of a foamed part with filler (both fillers and glass fiber) to the non-foamed part with filler:

$$R_{ff} = \frac{2t^3}{h^3} + \frac{6t(h-t)^2}{h^3} + (1 - R_{gf})\left(\frac{h-2t}{h}\right)^3 \tag{6.8}$$

The input data required for this calculation are also only the whole part weight reduction R_w, skin thickness t, and thickness of the whole specimen h.

The flexural strength data of filled PP are shown in Table 6.5. The results of the predicted flexural strength loss versus weight reduction percentage are acceptable with the error percentage of prediction range from 1.0% to 2.3%.

Table 6.5 *Flexural Strength Ratio Comparisons: 20% Talc Filled PP*

	Sample		
	1	**2**	**3**
Weight reduction, %	5	10	15
Measured R_f	0.960	0.920	0.880
Predicted R_f	0.944	0.914	0.884
% Error of prediction	−1.7	−0.7	0.4

The predicted fiber-filled PBT flexural strength also matches the actual data well with the error percentage less than 8.5% (see Table 6.6). It has been found in several investigations that the flexural strength of microcellular molded parts with glass fiber-reinforced material is even higher than in a solid part; it was also shown in the paper by Spindler. This is only true for the microcellular parts made with glass fiber-filled material. It is believed that fine cells help in disorientation of the glass fibers.

Figure 6.16 shows the results of modeling for flexural strength loss versus whole part weight reduction with the same materials and conditions as the ones in Fig. 6.15. The weight reduction percentage should be controlled to less than 25% for unfilled material and to less than 30% for glass fiber-reinforced material if no more than 20% flexural strength loss is allowed.

Table 6.6 *Flexural Strength Ratio Comparisons: 30% Glass Fiber Filled PBT*

	Sample		
	1	**2**	**3**
Weight reduction, %	4	10	15
Measured R_f	0.910	0.850	0.790
Predicted R_f	0.912	0.883	0.857
% Error of prediction	0.2	3.9	8.5

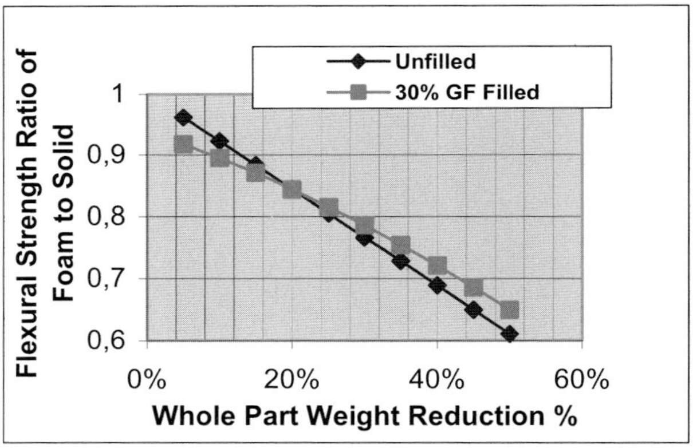

Figure 6.16 *Flexural strength loss versus whole part weight reduction % for filled and unfilled materials*

6.6.3 Impact Strength Model and Results

Impact strength modeling is very complicated because it is strongly related to a structure-property and density relationship. A fundamental model based on weight reduction, skin to core ratio, and cell density can be written as:

$$R_i = \sqrt[3]{\frac{N_m}{N_s}}(1 - R_n)^4\left(1 - \frac{2t}{h}\right)^{-1} \tag{6.9}$$

Where

R_i is the impact strength ratio of foamed part to solid part.

R_n is the real weight reduction ratio of the core material. It can be calculated either with Eq. 6.3 for unfilled materials or with Eq. 6.4 for filled materials.

N_m is the cell density for the microcellular part which is calculated from the foamed core section.

N_s is the reference cell density for standard foam and will be calculated with an average cell size of 100 microns.

The results of modeling are acceptable for filled materials such as PP with 20% talc and PBT with 30% glass fiber as shown in Fig. 6.17.

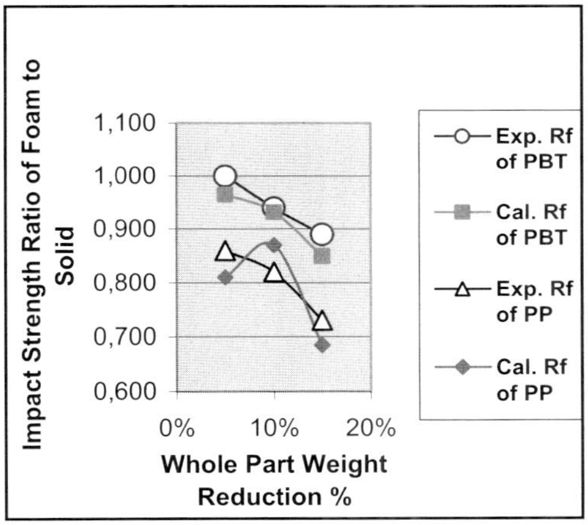

Figure 6.17 *Impact strength ratio of solid to foam versus whole part weight reduction % for filled materials*

6.6.4 Modeling Summary

Following are some observations from the physical property modeling.

1. The skin thickness is not an important factor for the tensile strength loss versus weight reduction. But it is a critical factor for the flexural strength of a microcellular part.
2. All models can be used for general predictions of the mechanical property loss versus weight reduction percentage for any material. The prediction error percentage is less than 10% (and usually less than 5%) when the foamed part has microcellular structure, i.e., the average cell size is equal to or less than 100 microns.

3. Generally, filled materials will have less tensile, impact, and flexural strength loss than the unfilled material at the same weight reduction percentage and skin thickness.
4. For reasonable mechanical property loss, such as less than 20%, the recommended maximum weight reduction ranges from 10 to 25%. Fine cells and uniform cell distribution are critical factors for high-impact strength and are very helpful for increasing tensile and flexural strength as well.

References

1 Kumar, V., Vander Wel, M.M., Paper presented at ANTEC 1991, Montreal, Quebec, Canada, May 1991

2 Shutov, F.A., Handbook of Polymer Foams and Foam Technology, Chapter 3 (D. Klempner and K.C. Frisch, Eds.), Hanser Publishers München (1991)

3 Spindler R., Paper presented at SME Conference, Seattle, WA, May 2001

4 Semerdjiev, S., *Introduction to Structural Foam*, SPE, pp. 6–7 (1982)

7 Moldflow Modeling of Microcellular Molding

7.1 Introduction

Since late 2000, Sejin Han, R. Zheng, and Peter Kennedy of Moldflow Corp. (a holding company of Moldflow Pty. Ltd., Australia) and Jingyi Xu and Levi Kishbaugh of Trexel, Inc. have worked together to study and develop a simulation model for the microcellular injection molding process. For the simulation, the microcellular molding process was defined as a supercritical fluid such as carbon dioxide or nitrogen being mixed with molten polymer to form a single-phase solution and then injecting the single-phase solution into a mold. The model simulates the development of cells in the melt during injection molding. The effects of cell growth on material properties and flow have been investigated. Some simulation results such as melt pressure and final cell size distribution are compared with experimental results.

The microcellular foaming process involves the introduction of carbon dioxide or nitrogen as a supercritical fluid (SCF) into a polymer melt to produce a microcellular foam [1, 2]. Such foams have cell sizes ranging from 5-100 microns and exhibit excellent mechanical properties. The amount of material that is injected into the mold is less than the part volume. Nucleation of cells and the subsequent cell growth fill the rest of the cavity.

The results of this study have been previously reported [3, 4]. The latest reported results show good agreement with the experimental results in terms of cavity pressure and the final bubble-size distribution.

7.2 Theory

In the microcellular foam molding process, insufficient material is injected to fill the mold. After injection, the nozzle is sealed and the rest of the cavity is filled by the nucleation and growth of cells. The key element in this process is therefore a cell growth model. This in turn needs to account for the diffusion of gas from the single-phase solution to the cell. Finally, the cell growth model needs to be coupled to the macrophase pressure distribution of the single-phase material in which it is embedded.

7.2.1 Cell Growth

To simplify the modeling, uniform nucleation throughout the melt and a known value for the number of nuclei per unit mass are assumed. To determine the growth of cells, the cell model proposed by Amon and Denson [5] is used. In the unit cell model (Fig. 7.1), the foam is divided into spherical microscopic unit cells of equal and constant mass, each consisting of a liquid envelope (or shell) and a concentric spherical gas bubble. In Fig. 5.1, R is the bubble radius and S is the outer radius of the cell. It is also assumed that the number of the cells equals the number of nuclei. Expansion of the bubble occurs due to two mechanisms: hydrodynamically controlled growth and diffusion-controlled growth.

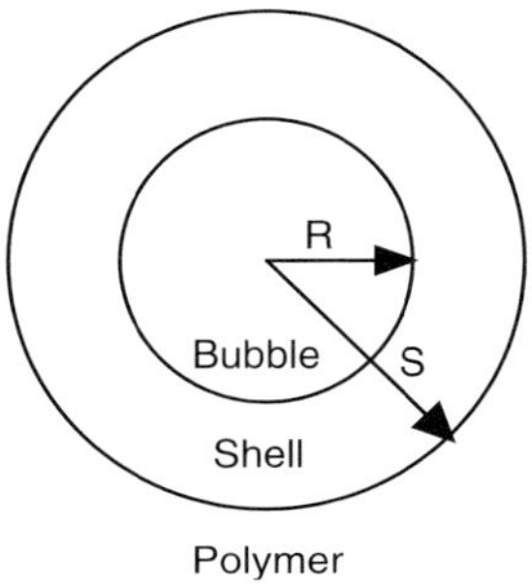

Figure 7.1 *Schematic diagram of a unit cell*

7.2.2 Hydrodynamic Growth Equation

The rate of change of the radius of the gas-melt interface, R_f, is given by [6]

$$4\eta\dot{R}/R = (P_g - P) - 2\sigma/R \tag{7.1}$$

where η is the melt viscosity, P_g is the gas pressure inside the cell, σ is the surface tension at the interface of the melt and the gas, and P_g is the pressure of the melt at the outer boundary of the cell. The internal gas pressure (P_g) is determined by diffusion while the pressure at the outer boundary of the cell $P_g)$ is determined by the macroscopic pressure equation governing the molding process.

7.2.3 Gas Diffusion Equation

The pressure inside the bubble (P_g) in Eq. (7.1) varies with bubble radius R. The relationship between P_g and R can be derived from the mass balance in the diffusion process of the gas dissolved in the melt envelop. The process is governed by Fick's law of diffusion:

$$\frac{\partial c}{\partial t} + v_r \frac{\partial c}{\partial r} = D\left[\frac{1}{r^2}\frac{\partial}{\partial r}\left(r^2 \frac{\partial c}{\partial r}\right)\right] \tag{7.2}$$

where D is the diffusion coefficient of the gas in the polymer melt, and c is the concentration (weight fraction) of the dissolved gas in the melt. Assuming that thermodynamic equilibrium is maintained at all time, the initial and boundary conditions are as follows. Initial condition is given by:

$$c(r,t=0)=c_0 \tag{7.3}$$

where c_0 is the initial gas concentration. The concentration at the bubble wall can be described by Henry's law,

$$c(R,t)=c_w(t)=\kappa_h P_g(t) \tag{7.4}$$

where κ_h is the Henry's law constant (solubility). For a given gas, its solubility in different polymers is about the same, but different gasses have considerably different values of solubility. In the absence of experimental data, the following simple empirical equation can be used to estimate the solubility [7]:

$$\ln \kappa_h = -2.1 + 0.0074 T_{cr} \tag{7.5}$$

where T_{cr} is the critical temperature.

We also assume that

$$c(r=S)=c_0 \tag{7.6}$$

and

$$\left.\frac{\partial c}{\partial r}\right|_{r=S}=0 \tag{7.7}$$

We may further assume that the gas follows the ideal gas law,

$$P_g(t)=\left(1000R_gT/M_w\right)\rho_g(t)=A\rho_g(t)T \tag{7.8}$$

where R_g is the universal gas constant (8.31451 J.mol^{-1}.K^{-1}), T is the temperature (in K), M_w *is* the molecular weight of the gas, ρ_g is the density of the gas in the bubble, and $A=1000R_g/M_w$ is a constant for a given gas. At the bubble wall, the mass flux is given by

$$\frac{d}{dt}\left(\rho_g R^3\right)=3R^2\rho D\left(\frac{\partial c}{\partial r}\right)_{r=R} \tag{7.9}$$

To calculate the concentration gradient, we solve Eq. (7.2) first and then substitute the solution into the RHS of Eq. (7.9). Integrating Eq. (7.2) with respect to r gives

$$\frac{d}{dt}\int_R^S (c_0 - c)r^2 dr = DR^2 \left(\frac{\partial c}{\partial r}\right)_{r=R} \tag{7.10}$$

Combining Eqs. (7.9) and (7.10) and integrating the resulting equation with respect to t, we get

$$3\rho \int_R^S \frac{c_0 - c}{c_0 - c_w} r^2 dr = \frac{\rho_g R^3 - \rho_{g0} R_0{}^3}{c_0 - c_w} \tag{7.11}$$

Following Rosner and Epstein [7], we assume a second order polynomial profile for the gas concentration,

$$\frac{c_0 - c}{c_0 - c_w} = \begin{cases} (1-\xi)^2 \ldots\ldots R \le r < S \\ 0 \ldots\ldots\ldots\ldots\ldots r \ge S \end{cases} \tag{7.12}$$

where $\xi = (r - R)/(S - R)$.

By substituting Eq. (7.12) into the LHS of Eq. (7.11), we can obtain an expression for S-R.

From the above equations, we obtain

$$\frac{d}{dt}\left(\frac{p_g R^3}{R_g T}\right) = \frac{6\rho^2 D \kappa_h R_g T (p_{g0} - p_g)^2 R^4}{p_g R^3 - p_{g0} R_0{}^3} \tag{7.13}$$

Eqs. (7.1) and (7.13) are to be solved to obtain information on bubble size and gas pressure inside bubbles.

7.2.4 Rheology of the Mixture

In order to describe the viscosity of the gas-polymer mixture, a modified Cross-WLF equation is proposed:

$$\eta(\dot{\gamma}, T, p, \phi) = \eta_0(T, p) f(\phi) \left[1 + \left(\frac{\eta_0(T, p)\dot{\gamma}}{\tau^*}\right)^{1-n}\right]^{(-1)} \tag{7.14}$$

where η_0 is the zero-shear viscosity which is a function of temperature and pressure, n and τ^* are model parameters, ϕ is the volume fraction of the gas, and f is a function of the volume fraction of the gas. The zero-shear viscosity is given in Eq. (7.15)

$$\eta_0(T, p, \phi) = D_1 \exp\left(-\frac{C_1(T - T_0)}{C_2 + (T - T_0)}\right) \tag{7.15}$$

with $T_0 = D_2 + D_3 p$ and D_1, D_2, and D_3 are model parameters.

In the injection molding process, the dissolved gas acts as an internal plasticizer, and the plasticizing effect of the gas can reduce the bulk viscosity. We use the following equation to describe this effect:

$$f = (1 - \phi)^{\alpha} \tag{7.16}$$

where α should be determined from experimental data. The volume fraction of the gas can be calculated by

$$\phi = \frac{4\pi R^3 / 3}{V_{shell} + 4\pi R^3 / 3} = \frac{4\pi R^3 / 3}{1/(\rho N_w) + 4\pi R^3 / 3} \tag{7.17}$$

7.2.5 Macroscopic Flow

To solve for the injection molding problems, the microscopic model of the cell should be coupled with the macroscopic flow through the fluid pressure around the cell.

In the macroscopic simulation, we treat the microscopic multi-phase fluid as a macroscopic single-phase fluid, thus all conventionally used governing equations, including the equations of continuity and momentum, as well as a constitutive equation, can be directly employed in the simulation [9]. The Hele-Shaw approximation can also be applied.

For mathematical convenience, we assume that the foaming stage starts after the injection nozzle is closed. Before the nozzle is closed, the mold filling analysis is therefore the same as for the conventional cases. After the nozzle is closed, the bubble expansion is the driving force behind the fluid flow. The pressure equation is given as follows:

$$\frac{\partial}{\partial t}\int_0^b \ln(\rho_{cell})dz - \nabla(S\nabla p) = 0 \tag{7.18}$$

with

$$S = \int_{-h/2}^{h/2} \frac{z^2}{\eta} dz \tag{7.19}$$

where $\nabla_{(x,y)}$ denotes the gradient operator with respect to the midplane, h is the thickness of the cavity, and z is the local thickness coordinate. Note that both the cell viscosity η and the cell density ρ_{cell} depend on the bubble size. The cell viscosity is given by Eq. (7.14), and the cell density is given by

$$\rho_{cell} = \frac{m_{cell}}{V_{shell} + 4\pi R^3 / 3} = \frac{3\rho}{3 + 4\pi R^3 N_m \rho} \tag{7.20}$$

7.2.6 Numerical Algorithm

The full system of governing equations consists of the coupled differential Eqs. (7.1), (7.13), and (7.18), and the energy equation (which is not given here for brevity). The primary unknowns to be solved are fluid pressure P, the gas pressure inside the bubble P_g, the fluid temperature T, and the bubble radius R. Equations (7.14) and (7.20) are used to update the viscosity and density respectively.

7.3 Experimental Work and Simulation

Moldings were made using Dow Chemical Styron 666D polystyrene material with 0.25% by weight of N_2. The mold is a rectangular plaque 279.4 mm long, 126.5 mm wide, and 2.55 mm thick. There is a 9.535 mm radius fillet on each of the four corners. The plaque is filled via a sprue at its center. The sprue is 74.5 mm long and has a tapered diameter of 4.11 mm at the nozzle and of 7.08 mm at the part (Fig. 7.2). Pressure transducers (Kistler 6157A) are located 25 mm from the sprue centrally along the longest axis and 126.5 mm away from the sprue. Data from the transducers was stored using an Engel CC100 machine control system.

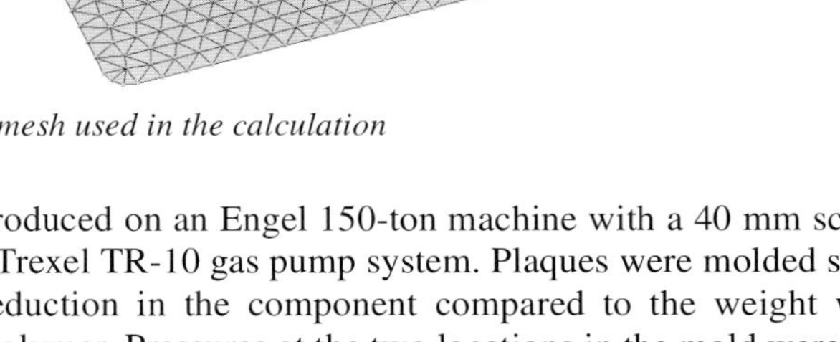

Figure 7.2 *Cavity mesh used in the calculation*

The parts were produced on an Engel 150-ton machine with a 40 mm screw. The machine was fitted with a Trexel TR-10 gas pump system. Plaques were molded such that there was a 15% weight reduction in the component compared to the weight when no gas was dissolved in the polymer. Pressures at the two locations in the mold were recorded, as were traces of ram displacement and hydraulic pressure.

Simulations were run using process conditions used to mold the samples. The gas-polymer solution was considered a single-phase solution during injection. Injection of material was stopped at 85% of the part volume and the nozzle was closed. Thereafter, it was assumed that the material expanded to fill the cavity due to cell growth. Cells were assumed to nucleate when the temperature and pressure was such that the gas precipitates from solution. Viscosity data were measured in a specially designed rheometer to allow measurement at varying levels of N_2. The measured viscosity data for 0.25% N_2 concentration are shown in Fig. 7.3. The data were fitted to the viscosity Eq. (7.14), and the fitting constants are shown in Table 7.1. The dependence of viscosity on bubble growth is not available experimentally, so an empirical value of $\alpha = 2$ was used in the current analysis.

The thermal properties and pvT data are based on solid resin (that is, without gas).

Table 7.1 *Viscosity Data Used for Simulation*

D_1 (Pa.s)	D_2 (K)	D_3 (K/Pa)	C_1
3.102e10	373.15	0	24.03
C_2(K)	τ^* (Pa)	n	α
51.6	2.839e4	0.1743	2

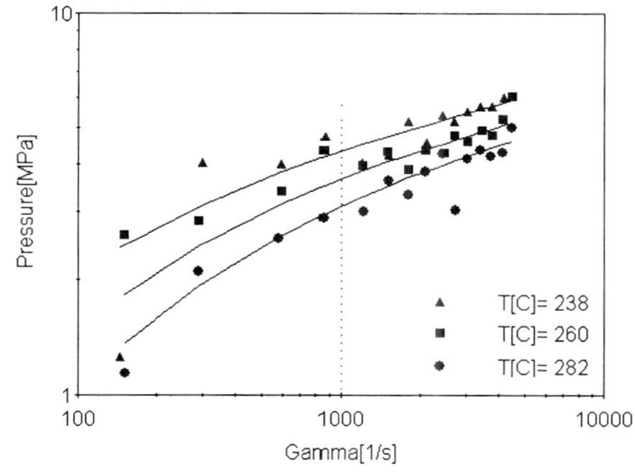

Figure 7.3 *Measured pressure vs. shear rate for viscosity*

The nucleation density is assumed to be 2×10^{11} nuclei per m^3, and $c_0 = 0.0025$, $\kappa_h = 8.4913 \times 10^{-10}$. Also, the diffusion coefficient is assumed to be 1.5×10^{-9}, and the initial bubble radius is at 1.0×10^{-6} m, and surface tension is assumed to be 0.05 (N/m). In reality, the cell density varies with location. In the current study, the highest cell density value among the measured values (which corresponds to the value at the farthest location from the sprue) is used.

Pressure results from the experiment and the calculations are shown in Fig. 7.4a and b. In this figure, "Calc" stands for the calculated results, and "Exp" for the experimental results. Predicted pressure at the 25 mm location (location 1: shown in Fig. 7.4a) is in good agreement with the measured value before the start of foaming. During foaming stage, the calculated pressure is generally lower than the measured pressure. The measured pressure at the 126.5 mm from the sprue (location 3: shown in Fig. 7.4b) rises almost as soon as the material reaches location 1. This may be due to increased pressure in the air and gas ahead of the melt – an effect that is not taken into account in the simulation. The calculated pressure at this location is generally in good agreement with the measured pressure once the pressure reaches the peak value.

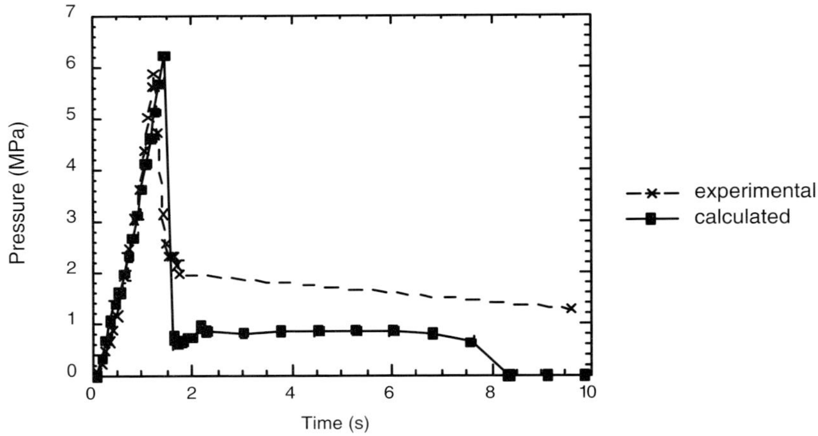

Figure 7.4a Pressure values at location 1

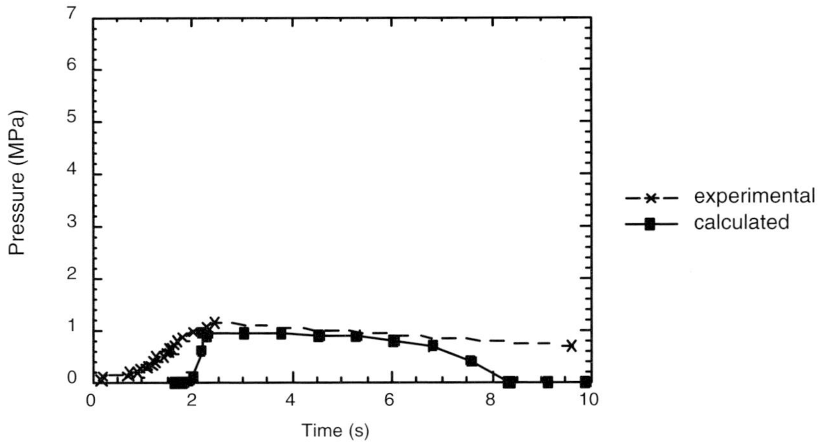

Figure 7.4b Pressure values at location 3

Calculated cell size is shown in Fig. 7.5. In this figure, the normalized thickness is from 0 to 1 (0 and 1 at the wall, and 0.5 at the core). The calculated size of cells increases with distance from the sprue which is due to the higher pressure encountered near the sprue reducing the rate of growth there. On the other hand, measured average cell radius decreases with distance from the sprue (for example, the values at locations 25 mm (location 1), 72.5 mm (location 2) and 126.5 mm (location 3) in the core region were found to be 50, 45, and 35 microns, respectively). The differences between the calculated and measured values are small near the sprue. However, the difference is rather significant in regions far from the sprue. Several reasons could account for this. One is the assumption of uniform nucleation. Experimental results show, however, that the nucleation is not uniform. The cell density far from the sprue is higher than the one near the sprue. The second reason could be the assumption of atmospheric pressure at the melt front. The experiment shows that the pressure at the melt front is higher than the atmospheric pressure because of the compressed air and gas at the unfilled region (Fig. 7.4b). The third reason could be inaccurate material data, especially the viscosity. The last reason could be the convection of bubble during foaming. This means that in an actual case, a small bubble upstream may move downstream during foaming, making the final bubble size downstream smaller also.

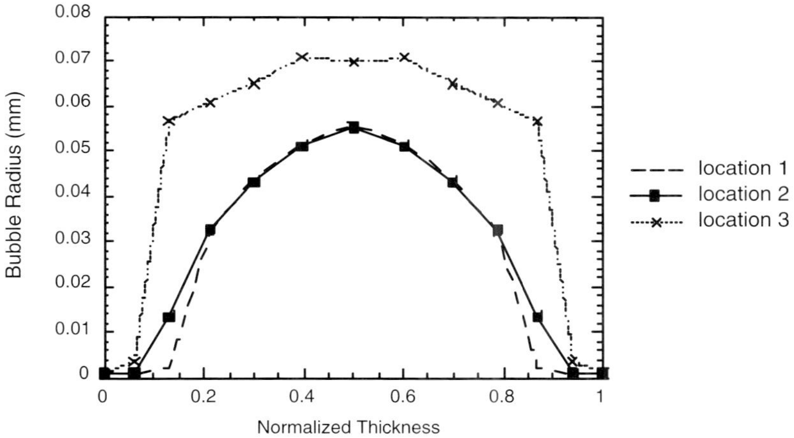

Figure 7.5 *Calculated cell size at locations 1, 2 and 3*

7.4 Summary

The model developed in this study can simulate the microcellular foam molding process reasonably well. Cavity pressure and the final bubble-size distribution are reasonably predicted with this model. Further work on the material data and modeling technique will be done in the future to further improve the accuracy of the prediction.

References

1. Martini-Vedensky, J.E. et al., US Patent No. 4 473 665 (1984)
2. Blizard, K., et al., US Patent No. 6 294 115 (2001)
3. Zheng, R., et al., Paper presented at ANTEC 2002, Dallas, TX, May 2002
4. Han, S. et al., Paper presented at ANTEC 2003, San Francisco, CA, May 2003
5. Amon, M., Denson, C.D., *Polym. Eng. Sci.* (1984) pp. 1026–1034
6. Eisenberg, P., Tulin, M.P., in Handbook of Fluid Dynamics, V. Streeter (Ed.) (1961) McGraw Hill, New York
7. Van Kreevlan, D.W., *Properties of Polymers: Their Estimation and Correlation with Chemical Structure,* (1976) Elsevier Scientific Publishing, Amsterdam
8. Rosner, D.E., Epstein, M., *Chem. Eng. Sci.* (1972) pp. 69–88
9. Kennedy, P., *Flow Analysis of Injection Molds,* (1995) Hanser Publishers, Munich

8 Microcellular Molding: Case Studies

8.1 Background

MuCell microcellular molding has been successfully performed on hundreds of injection molded parts with scores of parts now commercially produced for use in a wide variety of applications. Many injection molding components with high technical demands on part performance and/or processing benefit considerably from the MuCell process. Applied as a core plastics processing technology, the MuCell microcellular foam injection molding process is providing economic benefits across a broad component landscape including thin-wall parts requiring tight tolerances and dimensional stability; components with large differences in thickness and diameters; painted components; in-mold decorated and labeled components using fabrics and films; insert-molded parts; and components with high flow length to wall thickness ratios. The ability to more consistently mold thinner, flatter, dimensionally improved parts with the help of the microcellular foam process benefits all types of automotive and industrial end-users.

In the following pages, several case studies of parts successfully molded with the MuCell process are presented.

8.2 Mass Air Flow Sensor

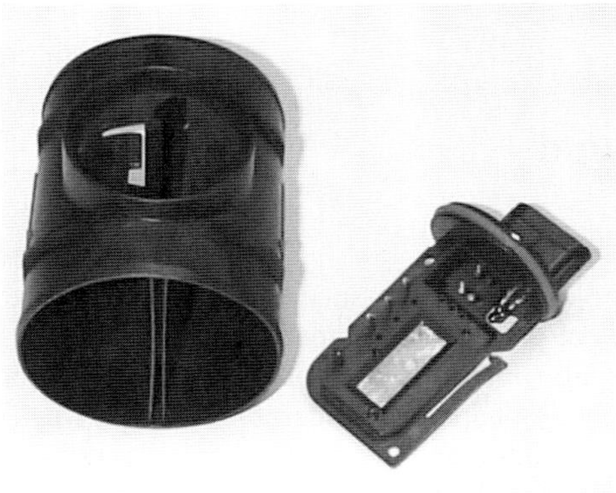

OEM:	Delphi Automotive Systems
Material:	Celanex 6500 30% glass/mineral-filled PBT

Delphi Automotive Systems has specified the MuCell® microcellular foam injection molding process for a high-volume mass air flow meter used to maintain the proper air/fuel ratio of the intake mixture to the engine. The air meter, or sensor, is an integral part of the advanced engine management systems used in today's vehicles. The accurate measurement of air flowing into the engine is used to determine the precise quantity of fuel needed to optimize engine performance and emissions.

The air meter provides this information by placing sensing elements in the air stream. Meter performance is enhanced by making the air flow around the sensor as consistent as possible and accordingly, the plastic components that houses the sensor and electronics must be extremely consistent.

Dimensional consistency is particularly important in the most recent designs by Delphi Automotive Systems. These have 1) an insert-molded plastic carrier that holds the electronic and sensing elements and 2) flow tubes that are mated using an all-plastic slide and latch system. The microcellular foam injection molding process significantly reduces the warpage caused by the uneven shrinkage observed in conventional solid injection molding and improves the meter's consistency and ease of assembly. The carrier and flow tube are both microcellular molded with a 30% glass fiber and mineral-filled PBT.

The next figure shows the improvement in dimensional stability possible with MuCell molding. The conventionally molded solid part was two to five degrees off parallel when comparing two parallel surfaces separated by a stem. The same part molded using the MuCell process was less than one-half degree off parallel.

< 0.5 degree off square | 2 - 5 degrees off square

MuCell Microcellular Foam Part | Solid Part

The next figure shows the reduced warpage in the bifurcator of the air flow sensor that is gained using the MuCell process versus conventional solid molding. With the reduced warpage, the part molded using the MuCell process assembles easily and the inserted o-ring centers naturally. This is not true for the solid molded part.

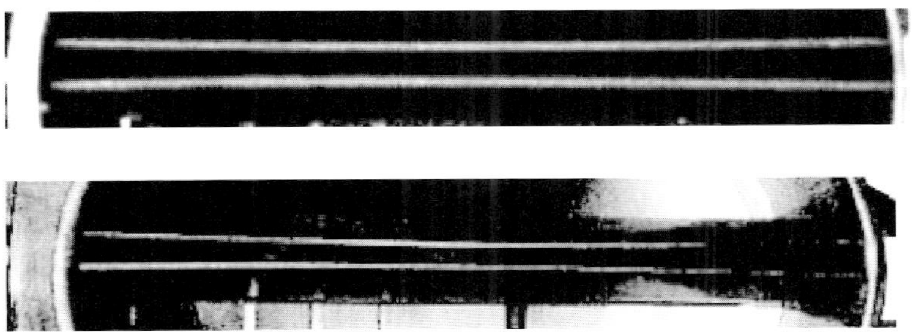

MuCell Microcellular Foam Part

Solid Part

Testing confirmed that the microcellular foam mass air flow sensor provides improved dimensional stability of the components, including critical sealing and press fit geometry, versus a conventionally molded meter from -40 to +200 °F, the operating temperature range of the meter. In addition, implementation of the microcellular foam injection molding process eliminated the cost of numerous tooling modifications typically needed to profile the mold to achieve proper dimensions.

The lower cavity pressures and mold temperatures of MuCell® microcellular foam molding are especially beneficial during insert molding of the "carrier" portion of the air meter.

This MuCell part was a 2002 SPE Automotive Innovation Award Finalist.

8.3 Automotive Fasteners and Door Latches

OEM:	Southco
Material:	Various

Southco, which became a Trexel licensee in 1999, is a 54-year old privately-held company whose parent has been in business for over 100 years. Southco is a global organization supplying a broad range of products to a number of industrial and automotive industries.

Southco was attracted to the MuCell process to take advantage of the potential material and cycle-time savings. After implementing the process though, Southco has learned that its customers are far more interested in the other benefits it can provide, such as improved product quality, the elimination of warpage and sink marks, and the improved dimensional stability without adversely affecting product performance.

Southco has at least ten parts in commercial production with MuCell molding today. Four of the parts are described below.

- Push-to-close Part: This part is made of glass-filled nylon. The MuCell process improved Southco's manufacturing performance by 25% through improved torque strength and better dimensional control.
- Cam for Push-to-close Part: By producing this cam with the MuCell process, warping was eliminated, which minimized rejections. Southco has run 80,000 to 90,000 of these parts so far.
- Draw Latch Link (2 sizes): These parts, which are made of glass-filled nylon, help to keep latches closed. Southco always had problems with this part due to warpage and sink marks because of the part's thick cross section. The MuCell® process allowed Southco to eliminate both problems, thus improving the part's quality and minimizing rejects. Due to the improved appearance of the MuCell molded part it is also more esthetically pleasing now.

8.4 Automotive Interior Trim

Molder:	Valeo
Material:	Various

Valeo is an independent industrial group fully focused on the design, production and sale of components, integrated systems, and modules for cars and trucks and ranks among the world's top automotive suppliers. The Valeo group of companies has 162 plants, 49 R&D centers, and employs 78,000 people in 25 countries.

Valeo has purchased multiple MuCell® machines focusing on interior trim applications such as the panel fascia shown above which uses an insert film molding process. The MuCell process lends itself especially well to this type of process due to the lower resin viscosity and thus reduced cavity pressure or melt temperature.

This application was a 2002 SPE Automotive Innovation Award Finalist.

Microcellular molding benefits for the fascia application shown are:

- Excellent processability with insert film molding
- Elimination of sink marks
- Improved dimensional stability
- Clamp tonnage reduced from 250 tons to 75 tons
- Part weight reduced 10%

8.5 Air Distribution Manifold

Molder:	Engel
Material:	20% talc-filled PP

An air distribution manifold was produced on a 900 ton Engel machine during the Interplas show in Birmingham, England (October 4-7, 1999). Both a cycle time and material weight reduction were demonstrated. Although these results are impressive, it is believed that with some mold modifications, the MuCell benefits would be even greater.

Demonstrated microcellular molding benefits are:

- 10% weight reduction
- 20% cycle time reduction

8.6 Engine Intake Manifold Gasket

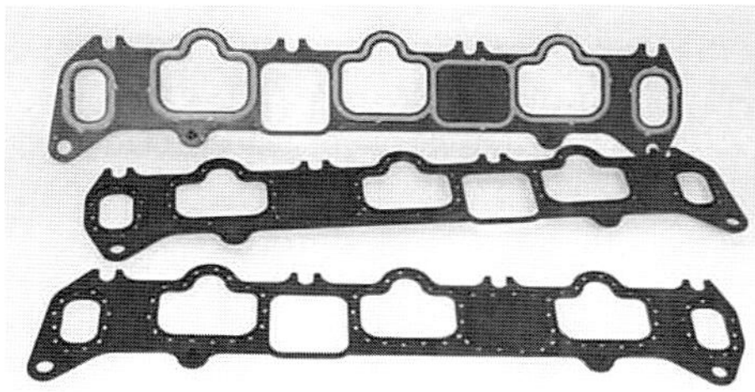

Material:	DuPont 33% glass fiber-reinforced nylon 6/6

An air intake manifold gasket was molded with a tremendous reduction in clamp tonnage, cycle time, and various weight reductions. The 0.116 in thick gasket was subsequently overmolded with thermoset silicone.

Microcellular molding benefits are:

- 5 to 20 % weight reduction
- Clamp tonnage reduced from 150 tons to 40 tons
- Cycle time reduced from 18 s to 6 s

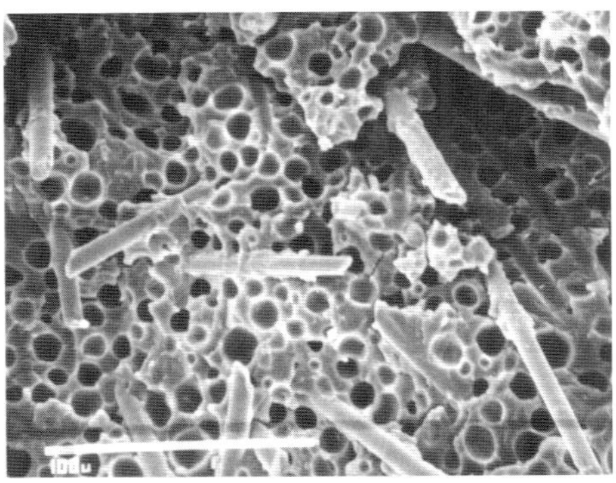

8.7 Automotive Mirror Bracket Support

Material:	25% mineral/15% glass fiber-filled nylon 6/6

The rear view mirror bracket demonstrated both a 50% cycle time reduction and up to a 28% reduction in weight.

Microcellular molding benefits are:

- Cycle time required to achieve dimensional stability reduced from 50 s to 24 s
- 28% weight reduction
- Passes physical property tests

8.8 PC/ABS In-mold Decoration – Automotive A-Pillar

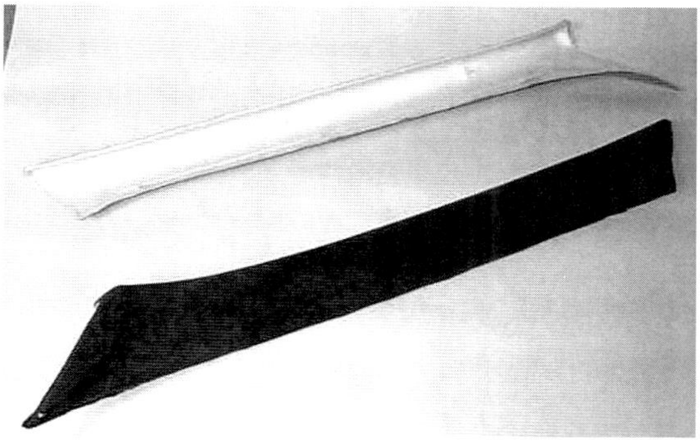

MuCell molding technology is used to reduce injection pressures and melt temperatures to improve the quality of in-mold decorated parts. MuCell material was injected onto a fabric.

Microcellular molding benefits are:

- Sequential valve gates removed and only center gate used
- Melt temperature reduced from 271 °C to 249 °C
- 5 to 10 % weight reduction
- Injection pressure reduced up to 28%
- No bleed-through due to low cavity pressures and melt temperatures

8.9 Automotive Fuse Box

Material:	25% glass-filled PP copolymer

This automotive fuse box with a living hinge has varying wall thicknesses with the thinnest section only 0.5 mm. A scanning electron microscopy (SEM) study of the part showed an average cell size of 50 microns.

Microcellular molding benefits are:

- 17% weight reduction
- Reduction or elimination of hold pressure
- Reduction or elimination of hold time
- Process temperature reduction of 56 °C

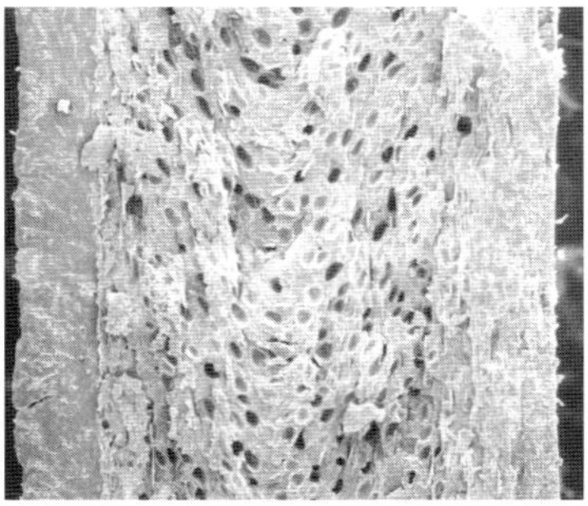

8.10 Rear Air Blower Base Connector

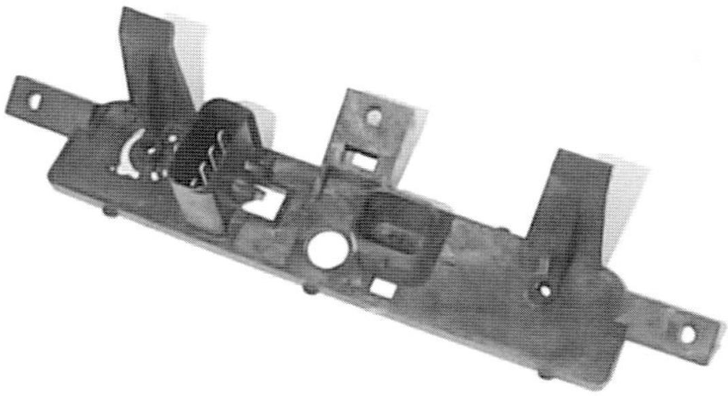

Material:	Glass fiber- and mineral-filled nylon 6/66

Microcellular molding benefits are:

- Flatness variance reduced from 0.06 in. to 0.008 in.
- 12% cycle time reduction
- 10% weight reduction
- 90% clamp tonnage reduction

8.11 Cargo AIP System

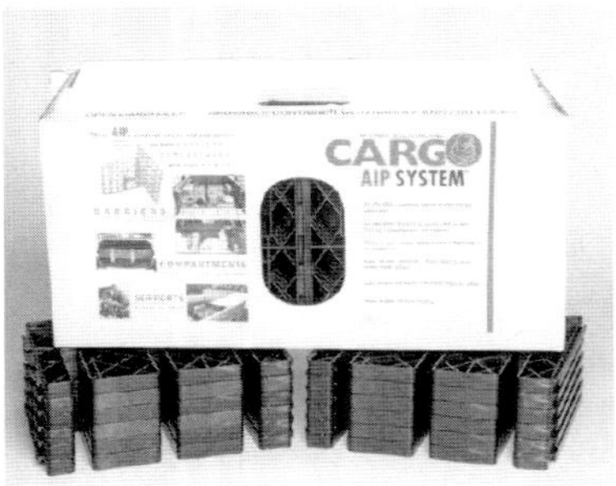

Molder:	Polymer Concepts
Material:	40% talc-filled PP

Microcellular molding benefits are:

- 8% weight reduction
- 17% cycle time reduction
- 75% clamp tonnage reduction
- Part warpage elimination

8.12 Fan Motor Housing

Material:	25% talc-filled polypropylene copolymer

Microcellular molding benefits in 0.5mm thick thin-wall application are:

- 15% weight reduction
- Eliminated sink marks
- Eliminated hold pressure and time
- 30% injection pressure reduction

8.13 Printer Components

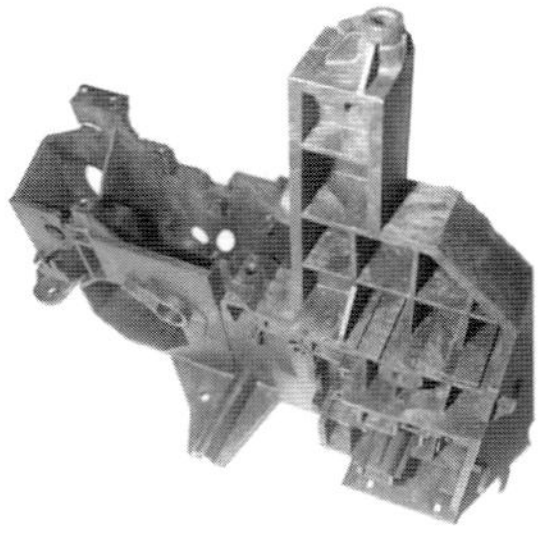

OEM:	Hewlett-Packard
Molder:	Avaplas, Ltd, Singapore
Material:	Various

Avaplas is a major molder for Hewlett-Packard (HP) and supplies components servicing industries in the printing and imaging, computer peripherals, telecommunications, semi-conductor, and customer electronics. Of their fifty-plus injection molding machines, Avaplas has five MuCell machines running commercial production on a series of internal structural components for HP printers.

Avaplas is running the MuCell technology on demanding structural components achieving excellent dimensional control at a significant cost advantage over conventional molding.

As an example of the benefits provided by the microcellular molding process, a lower paper guide has a warpage of only 0.429 mm with the MuCell process versus 0.807 mm with solid molding. The part is shown in the first figure below and the warpage across the part is shown in the next figure.

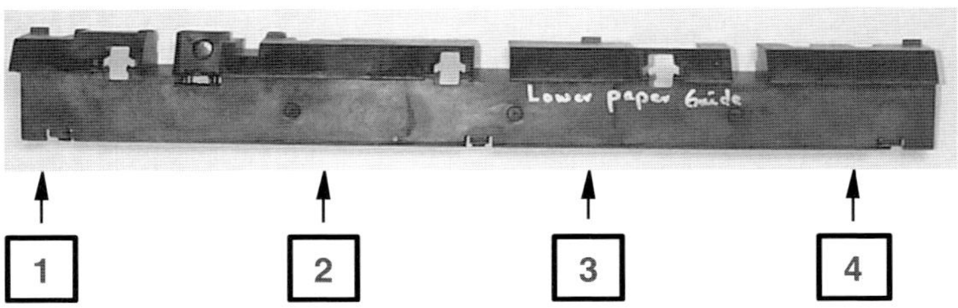

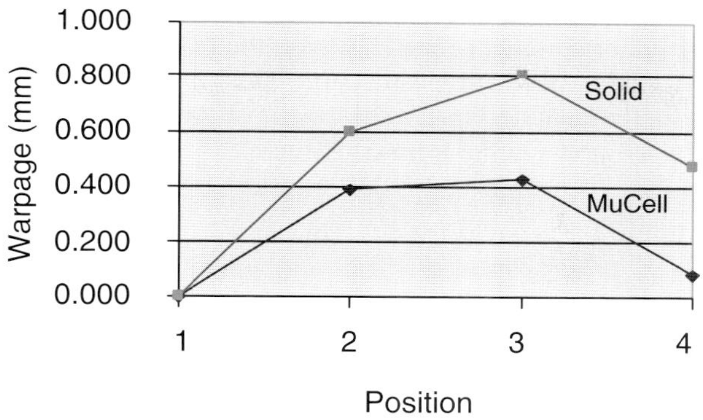

Other microcellular molding benefits in the lower paper guide are:

- Part stresses are eliminated
- Production costs reduced 4%
- Costly mold iterations to obtain flat parts eliminated

Another printer part demonstrates additional advantages of the MuCell microcellular molding process. The bonnet for a spittoon (see figure below) in an inkjet printer must meet tight dimensional specifications.

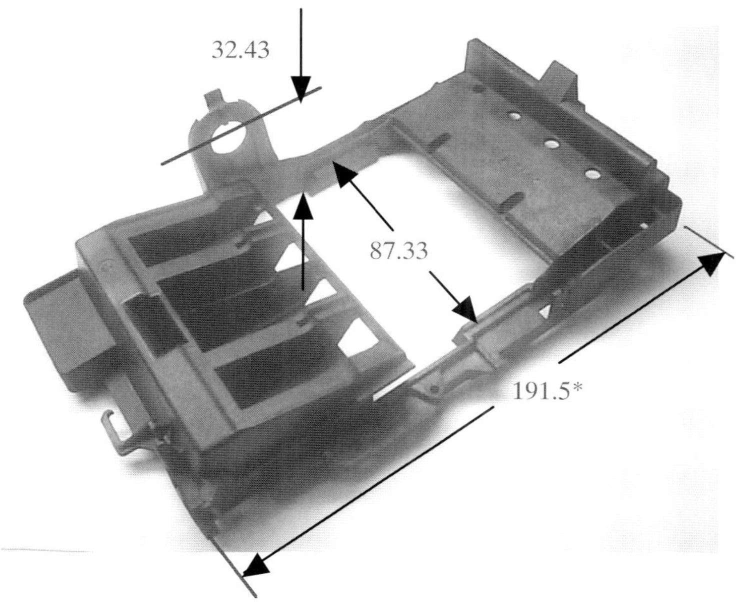

The reject rate in the solid process was unacceptable and was forcing HP to consider returning to metal structural components. However, the microcellular molding process has resolved the dimensional issues with plastic parts. The MuCell process provides parts that meet dimensional requirements without excessive costly tool iterations and improves the dimensional stability as shown in the following figure.

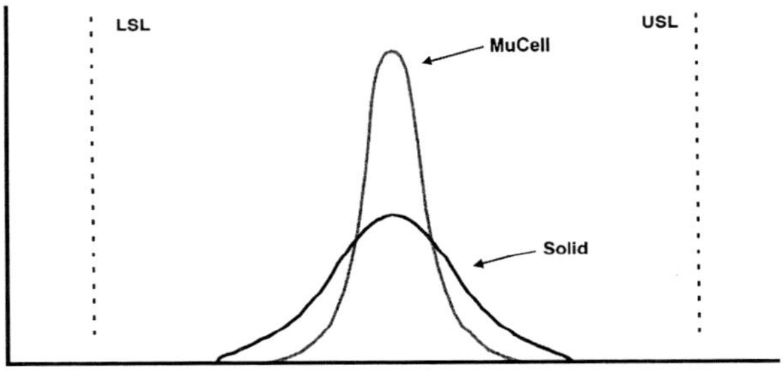

8.14 Precision Molded Parts

Molder:	Empire Precision Plastics
Material:	Various

Empire Precision Plastics (Rochester, NY, USA) specializes in high-volume, high-tolerance injection molded parts. Its customers are from the automotive, business equipment, electronics, consumer, and industrial sectors. Each of the parts Empire presently manufactures using the MuCell process had serious dimensional and performance obstacles to overcome when molded using conventional solid molding. Using the MuCell process, Empire prevailed.

In addition, the MuCell process works well with Empire's aluminum tooling. Empire uses aluminum tooling to improve thermal conductivity and the lower injection pressures of the MuCell process reduce the tool stress and thus increase tool lifetime.

On a level sensor made of HDPE, Empire reduced part weight by 10% and cycle time by 20%; on an industrial ice machine float made of PP, Empire reduced part weight by 10% and cycle time by 30%; and on glass-filled polyphenylene sulfide (PPS) valve parts (shown above), Empire reduced sink mark depth by 80%, part weight by 5%, and cycle time by 20%.

8.15 Electrical Switches

OEM:	Gerard Industries, Australia
Material:	Various

Gerard Industries, a major supplier of electrical accessories doing business in over 25 countries on 5 continents has several MuCell commercial applications and multiple MuCell production machines. The switch shown above is one of several MuCell parts being manufactured by Gerard Industries.

Microcellular molding benefits are:

- Weight reduction
- Cycle time reduction
- Clamp tonnage reduction
- Dimensional stability

8.16 Electronic Control Module (ECM) Connector

Material:	Glass fiber-reinforced PBT

Microcellular molding benefits are:

- Pin retention force and standard deviation (S.D.) improved
  - Solid molding: 36.7 N (4.1 N S.D.)
  - MuCell molding: 37.8 N (2.9 N S.D.)
- 10% weight reduction
- Elimination of sink marks
- Improved flatness
- Increased assembly yields
- Improved sealing

8.17 Power Distribution Center Plate

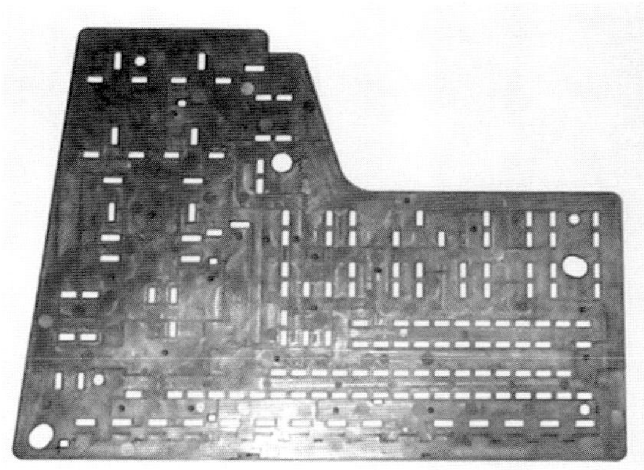

Microcellular molding benefits are:

- Improved flatness- elimination of bow
- Dimensional stability

8.18 Data Communications Box

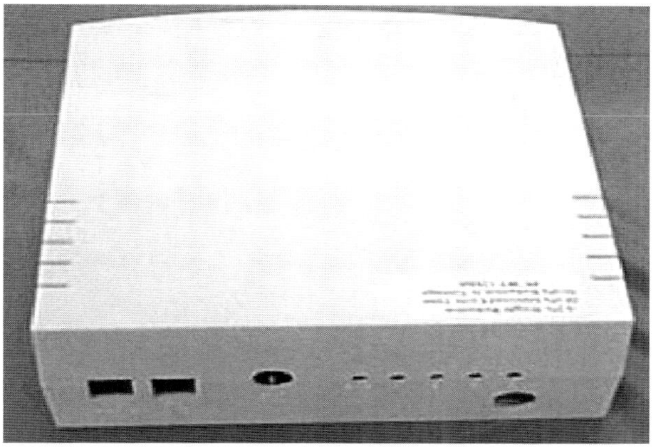

Microcellular molding benefits are:

- 7% weight reduction
- 28% cycle time reduction
- 50% clamp tonnage reduction

8.19 Electrical Box

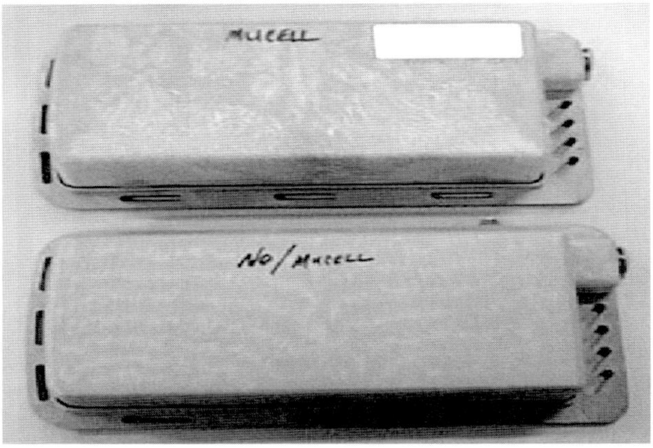

Material:	Clay-filled PP

Microcellular molding benefits are:

- 5% weight reduction
- 19% cycle time reduction
- 50% clamp tonnage reduction

8.20 Electrical Connector

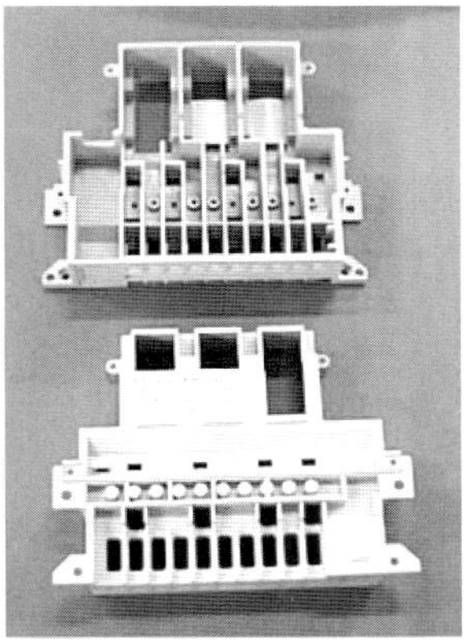

Microcellular molding benefits are:

- 18% weight reduction
- 25% cycle time reduction
- 80% clamp tonnage reduction

8.21 Wire Spool

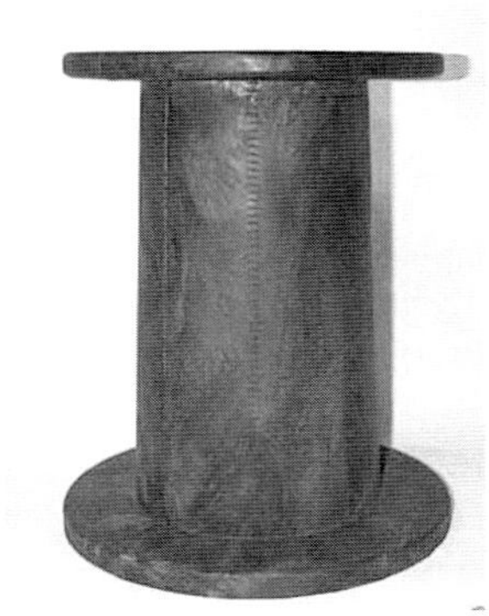

Material:	ABS

Microcellular molding benefits in 0.5 mm thick thin-wall application are:

- Up to 30% weight reduction
  - 1.6 kg solid part
  - 1.1 kg MuCell part
- 20% cycle time reduction
- Reduced clamp tonnage drastically

8.22 High-Wear Components

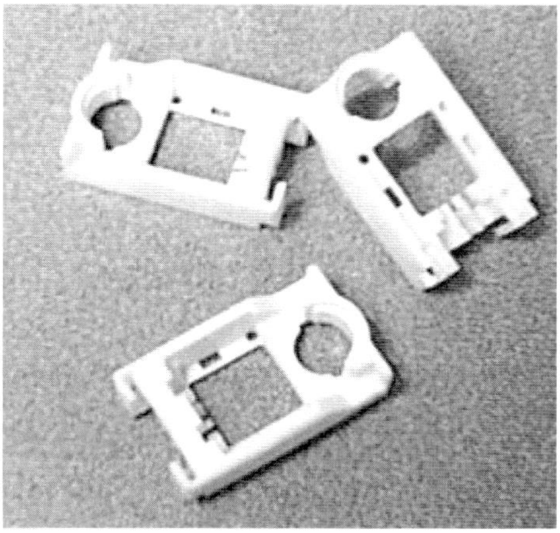

Material:	Acetal

Microcellular molding benefits are:

- 10 to 40% weight reduction
- 20% cycle time reduction
- Clamp tonnage reduced from 150 tons to 40 tons

8.23 Cable Ties

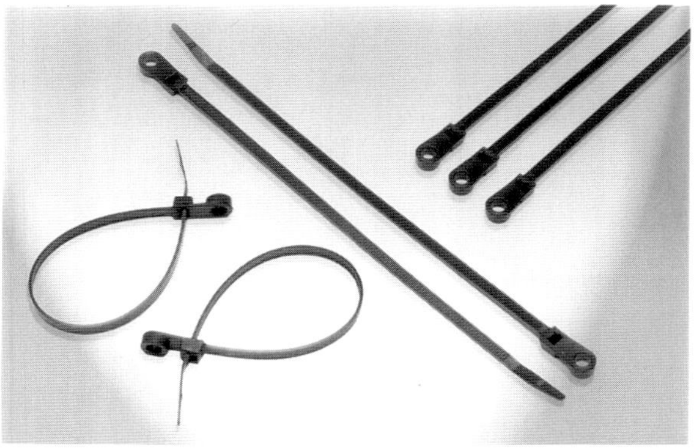

Material:	Nylon 6/6

Application studies on the use of nylon 6/6 in microcellular injection molding applications have produced the following results:

- 8 to 10% weight reduction at a thickness of 1.45 mm
- Average cell size = 20 microns
- Eliminated hold pressure and hold time
- 30% reduction in hydraulic injection pressure
- 30% reduction in clamp tonnage (3.95 tons/in.^{2} to 2.76 tons/in.^{2})

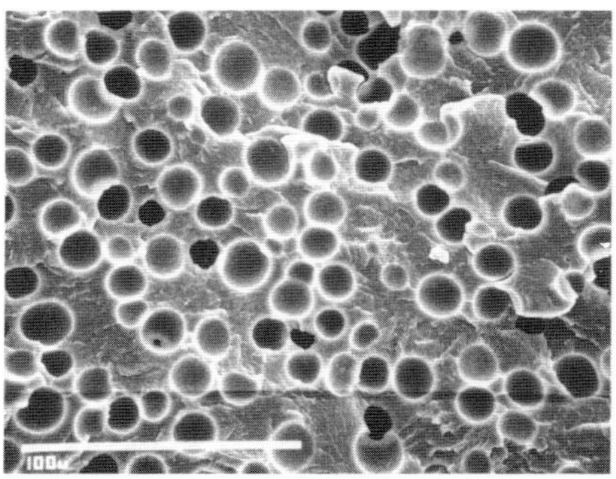

8.24 Medical Staple Gun

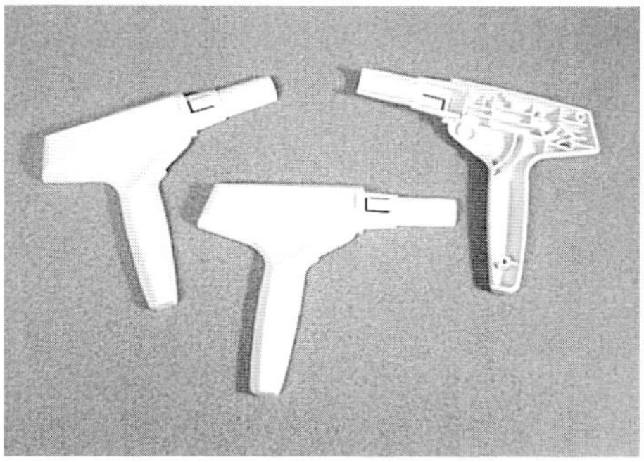

Material:	Radel PPS; polycarbonate (shown)

Microcellular molding benefits are:

- 30% weight reduction
- up to 50% cycle time reduction
- Clamp tonnage reduced from 120 tons to 15 tons
- Mold temperature reduced from 80 °C to 15 °C

8.25 Syringe Plunger Cap

Material:	Santoprene TPV

Microcellular molding benefits are:

- 23% weight reduction
- Reduced Shore A hardness
- Improved sealing capability
- Improved compression set
- No oil additives
- No fogging issues

8.26 Child Gate Panel

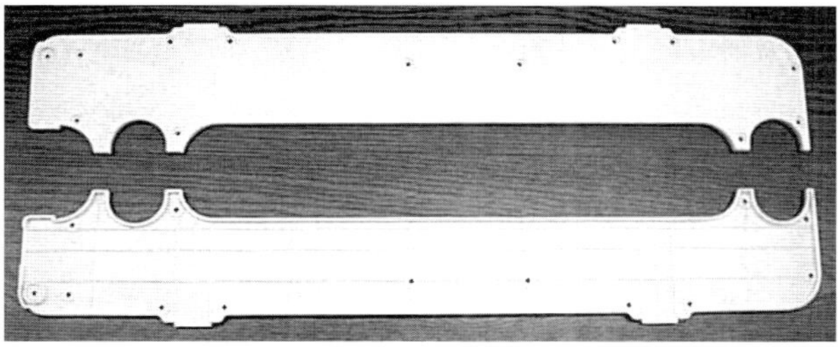

OEM:	SafetyFirst
Molder:	MarLee

Microcellular molding benefits are:

- 11% weight reduction
- 18% cycle time reduction
- Clamp tonnage reduced from 200 tons to 90 tons
- Injection pressure reduced from 170 bar to 100 bar
- Injection increased from 8.9 cm/s to 11.2 cm/s

8.27 Thin-Wall Container

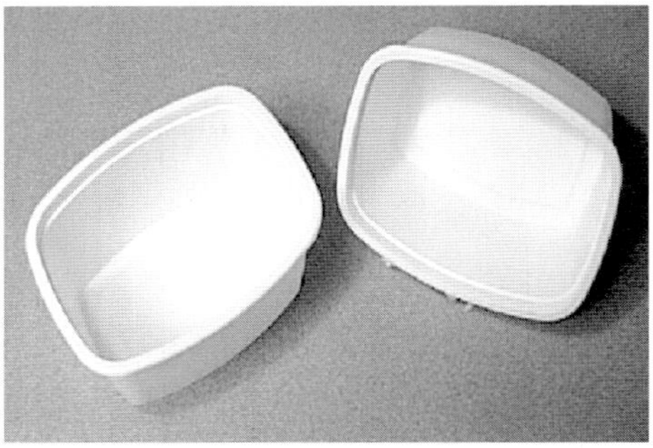

Material:	Polypropylene

Microcellular molding benefits in 0.5 mm thick thin-wall application are:

- 10% weight reduction
- 27% cycle time reduction
  - Eliminated hold time
  - Reduced cooling time to 0.1 s
- 36% clamp tonnage reduction
- 20% injection pressure reduction

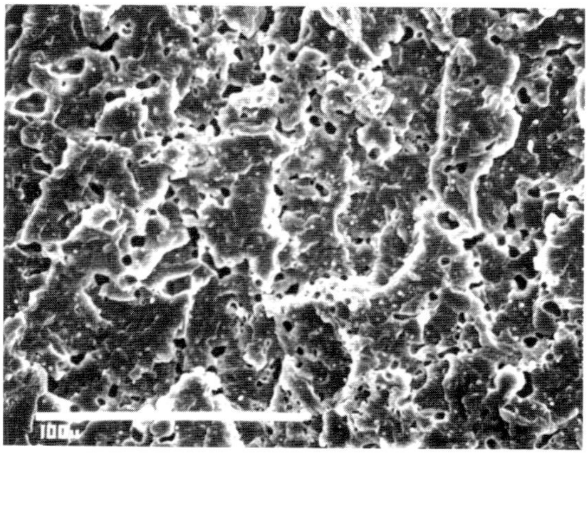

8.28 Bottle Cap

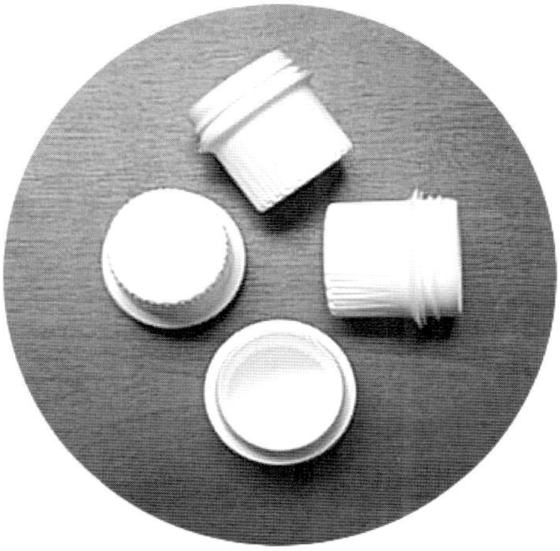

Material:	Polypropylene

Microcellular molding benefits in 0.5 mm thick thin-wall application are:

- Up to 20% weight reduction
- 25% cycle time reduction
- 50% reduction in master batch use
- Maintained critical performance
  - Hoop strength
  - Sealing lip performance

9 Microcellular Molding: Mold and Part Design

9.1 Background

MuCell microcellular molding is different from conventional solid molding in several fundamental ways that affect mold and part design.

- During injection, when using the MuCell process, the polymer melt is actually a single-phase solution of polymer and a supercritical fluid (nitrogen or carbon dioxide). This single-phase solution has a much lower viscosity than standard polymers.
- Microcellular cell development starts with a nucleation process, initiated by exposing the single-phase polymer melt solution to the abrupt pressure drop associated with mold filling.
- Cell growth occurs during mold filling and cooling. Microcellular molding requires neither a long hold time, nor a high hold pressure. The cavity pressure develops as the material expands to fill the mold. Cell size and uniformity are affected by the mold pressure and the cooling rate of the material.

Despite the process differences, microcellular molding can be practiced on most molds that work in conventional solid molding. Generally, the troubleshooting of a mold is similar for both the solid and the microcellular molding process. Although not intended to be a complete mold and part design guide, the following design recommendations will help a processor maximize the benefits offered by microcellular molding.

Mold and part designers should keep in mind the following general rules for microcellular molding:

- Allow for fast fill rates, especially when maximizing weight reduction by properly sizing runners, flow length, and vents.
- Provide a balanced flow to each gate and through the cavity-filling step.
- When possible, locate gates so as to help the molder develop a uniform fill rate and pressure drop across the part.
- Eliminate hot spots in the mold that could limit cycle time reduction.

9.2 Hot Runner and Sprue Design

In microcellular molding, cell nucleation occurs in the molding machine at the point where the pressure drops. The point of nucleation for a valve-gated hot runner or sprue is at the valve gate and thus is located at the part. This allows the molder to process for the part and not the runner system. Valve-gated hot runner and sprue systems have the same advantages.

As mentioned previously, it is necessary to keep the SCF laden polymer under pressure or it will foam in the barrel and hot runner system. This pressure on the polymer typically ranges from 100 to 170 bar but can be as high as 240 bar. This pressure will be applied to the material in front of the screw tip including the material in the hot sprue. Therefore, the valve gate must be able to withstand 240 bar of polymer pressure without leaking or drooling.

An additional requirement is that the hot runner system should be naturally flow balanced. This means the distance from the machine nozzle to each gate as well as the runner diameters should be the same. This allows for even filling of all cavities.

If possible, individual control over the valve gates will provide additional control of cavity balance. This is important because uneven filling of cavities will result in varying part weights across cavities and will minimize the ability to reduce part weight.

Finally, all hot sprues must be valve-gated. Non-valve-gated or thermal gates will not work with the MuCell process. If the system is not valve-gated, the polymer will foam in the hot sprue resulting in drool. Furthermore, thermal gates will not withstand the required polymer pressure of 240 bar.

9.3 Cold Sprue and Runner Design

The point of nucleation in a cold sprue is at the shutoff nozzle on the injection molding machine nozzle. Because the molder must optimize the microcellular molding process for part, sprues, and runners, the cold sprue and runner are often the cycle-limiting aspect with the MuCell process.

Knowing this limitation, cold sprue and runners can and have been used successfully in the microcellular molding process. When designing cold runner systems, the reduced viscosity of the polymer melt/SCF single-phase solution allows for a reduction in size of the sprue and runner. In addition, since the microcellular molding process does not rely on hold pressure to complete the filling process, further reductions in diameter are possible. Reducing the diameter of the sprue and runner also helps to reduce the cycle time limitation of cold sprues and runners.

In microcellular molding, the most common issue encountered with cold sprues is the sprue sticking in the sprue bushing and the runner system staying with the moving side of the mold. This is associated with two characteristics of the MuCell process: 1) the cell structure continues to force the sprue out against the bushing until the material cools sufficiently to withstand the pressure in the cells and 2) the foam acts as an insulator slowing the cooling effects in thick cross sections such as sprues.

In order to counter these effects, the mass of the sprue should be kept as low as possible. The mass of the sprue can be minimized by following these basic rules:

- Keep the sprue length to less than 50 mm.
- Make the sprue diameter at the top only slightly larger (0.8 – 1.5 mm) than the machine nozzle diameter.
- Use a draft angle of 2 – 6 degrees included angle.

The ability to reduce the sprue length is controlled by the part geometry and the length of nozzle penetration available on the molding machine. In cases where the sprue length cannot be reduced, where cooling times are short, or where the sprue is unusually thick, additional measures can be taken to control sprue sticking.

- Use a high thermal conductivity sprue bushing to improve cooling of the sprue
- Add a cooling circuit in or near the sprue bushing.
- Extend a cooling pin into the sprue bushing to reduce the material at the junction of the sprue and runner and to aid in heat removal. In this case, an ejector sleeve can be used to eject the runner from the cooling pin.

If possible, the sprue bushing should be produced from a high-conductivity material with a water line located within close proximity.

As the size of the sprue is reduced to optimize cycle time, gussets and radii can be used to enhance its strength. At the intersection of the sprue and runner, add a 2 to 3 mm radius or a small gusset. The gusset should be about 1.5 mm up the sprue and 5 mm along the runner. The gussets should be 2 to 3 mm thick.

Figure 9.1 illustrates some of the cold sprue design and cooling aspects. Cooling will be most effective if the water lines are placed in or as close as possible to the sprue bushing.

As with hot runner systems, cold runner systems must be naturally flow-balanced. This means that the distance from the sprue to the gate must be the same for all cavities and the runner diameters should be the same. This will result in the most uniform fill pattern and will allow for better control of weight reduction by minimizing cavity-to-cavity variations.

The runner should be sized to approximately the thickness of the part. This can be accomplished because of the reduced viscosity of the supercritical fluid-laden polymer and the elimination of the hold phase of the molding process. These smaller runners will reduce cycle time as they will cool more quickly and also provide better cell control. The improved cell structure control results from higher pressure in the runner system, which slows cell growth until the material enters the mold cavity.

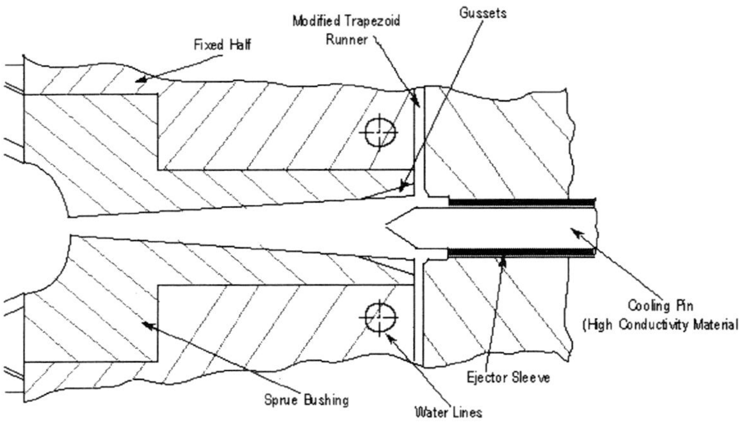

Figure 9.1 *Cold sprue design options*

9.4 Mold Balancing

Flow balance is always important in molding, but it is even more so in microcellular molding. Unlike with solid molding, microcellular molding uses very little cushion or packing pressures. Overpacking one cavity in solid molding will not significantly increase the density of the part. In microcellular molding however, this will create a higher-density part and a higher part weight, compared to other cavities.

It is easier to develop consistent microcellular parts with one- and two-cavity cold runner molds, and it is possible with four-cavity molds. However, fill patterns tend to change with 8 cavities or more because of the shear-induced heating that occurs within the cold runner system. Considerable effort is required to balance these molds for microcellular molding.

As with conventional molding, the 'H' pattern runner system for four-cavity molds or a radial runner design are best for balancing the mold cavity fill patterns.

Figure 9.2 *Balanced runner patterns*

To balance the runner system for microcellular molding:

- Examine short shots first. If necessary, adjust runner vents and runner sizes to balance.
- Gradually increase shot size and locate and eliminate any sources of imbalance. These imbalances include wall thickness variations, gate sizes and shapes, and trapped gas due to poor venting.
- Also, look for thermal imbalance by being particularly careful to avoid thermal differences that can result from water line loops.

The MeltFlipper™ runner designs developed by John Beaumont and colleagues at Penn State University-Behrend has also been demonstrated to improve cavity variation in MuCell microcellular molding [1, 2].

9.5 Three Plate Mold Designs

Three-plate molds for solid parts are typically designed with large sprues and runners to minimize pressure drop through the runner and maintain a flow channel through the runner until the end of hold pressure. This type of sprue and runner design sticks in the mold when demolding and typically limits cycle time reductions when using the microcellular molding process. A valve-gated hot runner is usually a better option for attaining maximum benefit with the microcellular molding process.

When three-plate molds are used, the sprue length should be less than 10 mm with good draft. This will eliminate issues with the sprue sticking in the bushing. Runners should be sized to allow easy filling of the mold cavity but, as with standard cold runner molds, the runner diameter can be reduced due to the lower viscosity as well as the elimination of the hold phase of the molding process.

The drops from the runner to the cavity should follow some of the same design consideration as the sprue, they should be as short as possible and radiused at the junction with the runner. There must be a cooling circuit in the runner plate and this should run close to the drops.

The sucker pins should be designed of high-thermal conductivity steel to assist in creating a higher strength skin around the pin. This helps in pulling the drops from the runner plate.

9.6 Gate Designs

The MuCell microcellular molding process does not require special gate designs. Molds have been run with all types of gate designs including sub-gates into ribs or ejector pins. The only gate type that often leads to issues is a curved tunnel gate.

When sizing gates, it should be noted that, as with runners, the gate does not need to be sized for hold pressure. This can allow for a smaller diameter gate. However, the shear generated with a smaller gate can create part performance problems.

Gate location controls the filling profile of the cavity. Locating the gate for the minimum flow length to thickness ratio and uniform fill pattern will result in the largest weight reduction and the lowest cavity pressure. If the gate is located such that the part does not fill evenly, the area to fill first will be at a higher density than the end of fill.

Another consideration regarding gating is wall thickness. The ideal situation is to have a uniform thickness across the part, i.e., less than 20% variation. However, this rarely occurs. In conventional molding, proper mold design has gates entering the thick part sections to reduce sink marks through better transfer of hold pressure. However, because microcellular molding does not require hold pressure and uses the internal gas pressure in the cells to provide the packing force, it can achieve higher weight reductions by gating into thin part sections and then flowing into thicker sections. If the thin section is at the end of fill, the cavity pressure will be higher and this will result in compression of the cell structure near the gate.

Typically there will be a different surface appearance around the gate compared to the bulk of the part. As the microcellular molding process is not recommended for appearance parts, this is most likely not a critical issue. However, keeping the gate off an appearance surface will move gate blush to a more acceptable area of the part.

In microcellular molding, gate placement should consider the flow factor (flow length to wall thickness ratio or L/t), particularly when maximum weight reduction is the goal of the molder. As Table 9.1 suggests, lower flow factors will help yield maximum weight reductions:

Table 9.1 *Correlation Between Flow Factor and Weight Reduction*

Flow factor (L/t)	Typical maximum weight reductions
50:1	+20%
100:1	15–20%
150:1	10–15%
200:1	5–10%

9.7 Venting

All injection molds require adequate venting. In traditional molding, the final filling and packing of the mold is accomplished under a condition of high pressure and low flow. This allows elimination of gas traps by forcing the gas out of the mold under these conditions. In solid molding, inadequate venting leads to visual defects such as short shots and burn marks and will also create hidden problems such as weak weld lines and higher molded-in stress patterns that can lead to premature failure of the part and dimensional problems during heat exposure. Furthermore, inadequate venting in solid molding results in a narrower processing window and longer cycle times.

The microcellular molding process completes the filling and packing of the mold under low pressure, which makes it difficult to force out trapped gas. The evolution of the gaseous blowing agent from the melt front and the higher injection speeds also make mold venting a critical area of concern in microcellular processing. The lower viscosity of the SCF laden polymer melt may slightly change the mold filling pattern but in most instances an area of gas trap in the solid process will be an area of gas trap for the microcellular molding process.

Proper venting of a mold is critical to producing a quality injection molded part. As in solid molding, inadequate venting with microcellular processing can result in chronic short shots. This typically occurs in a completely unvented perimeter area or a deep thin rib. A poorly vented mold can also reduce two important microcellular processing advantages: adding molded-in stresses and minimizing the attainable weight reduction. Correction of poor venting has been repeatedly demonstrated to increase weight reduction from about 5% to about 10%.

In microcellular processing, certain mold geometries should also be evaluated for additional venting. These geometries include thin ribs susceptible to back flow and blind pockets such as posts, locating tabs, snap fits, and butt welds.

9.7.1 Identifying Venting Issues

Venting problems can be identified in a number of ways. The easiest method to determine whether there is adequate perimeter venting for microcellular processing is to reduce clamp tonnage. As clamp tonnage is reduced, the mold vents better along the parting line. If the mold lacks perimeter venting or if the perimeter vents are damaged, the reduced clamp tonnage allows additional weight reduction to be achieved. Clamp tonnage can continually be reduced until flash forms at the parting line.

Perimeter venting can also be evaluated by placing a brass shim at each leader pin to keep the mold open 0.125 mm. Again, if the part can be molded at a significantly higher weight reduction than without the shims, inadequate perimeter venting is affecting the potential benefits of the microcellular process. After using either evaluation method, if perimeter venting is found to be inadequate, then the perimeter venting should be increased or reworked to maximize the microcellular processing window and benefits.

Venting problems in areas other than the perimeter are harder to diagnose. The best approach to identifying inadequate non-perimeter venting is to

- Mold the part using the microcellular process to about 95% full; do not try to pack out the last areas to fill. Note the injection pressures.
- Next, attempt to finish filling the part. Increase injection speed, injection pressure, hold time or hold pressure as needed. Note the injection pressures.
- Compare the pressures required to finish filling the part. If the part weight increases significantly before the final 5% of the part fills, these last to fill areas most likely have venting problems.

For example, a part run at Trexel had a weight reduction of 15% compared to solid when the part was slightly underfilled. However, when the part was completely filled, there was only a 4% weight reduction. This was an indication that excessive pressure was being used to fill the part thereby compressing cell structure and limiting the attainable weight reduction. Adding a local vent in the unfilled area resulted in an immediate reduction in the pressure required to fill the part and an immediate increase in weight reduction to 12%.

9.7.2 Proper Vent Design

When designing and building a mold for venting, there is very little difference between the recommendations for standard molding and for microcellular processing. The only significant difference is the depth of the vents at the end of fill.

The vent guidelines for microcellular processing are below. The guidelines for parting line vents are:

- Vent 40 to 50% of the perimeter. In new molds, vent 25% of the perimeter and after the first trial, increase vents in size or add new ones to critical areas.
- Vent widths should be increased to 5 to 12.5 mm.
- Vent depths should be toward the high end of material manufacturer recommendations for solid injection molding for areas that experience traditional filling pressures.
- Vent depths can be increased 50 to 75% or more for areas that are the last to fill. This should result in a 40 to 75% reduction in pressure at the end of flow.
- Vent land length should be minimized; no more than 0.5 to 1.0 mm long if possible.
- Relief channels: 1 to 2 mm. All relief channels should vent to atmosphere.

The guidelines for venting non-parting line areas are:

- Sprue Puller: 0.04 mm clearance or 0.05 mm flats on puller pin
- Runners: vent at every change of direction and if possible adjacent to the gate. Material supplier specifications should be used.

- Bosses and deep ribs: vent if deeper than 4 times the thickness
- Venting ejector pins: should have ground flats from 0.05 to 0.1 mm, for 3mm below the fit length of the pin. Typical pin clearances of 0.025 to 0.04 mm do not provide enough venting.
- If venting ejector pins are not practical in deep rib areas, it is recommended that the coring be fabricated (Lamella design) to provide vents at the base of ribs.
- Venting in slides: deep ribs and bosses in side actions should be vented where possible, utilizing venting pins and flats on core pins.
- Relief channels: 1 to 2 mm. All relief channels should vent to atmosphere.

9.8 Mold Cooling Designs

Mold cooling is the key to cycle time. When designing a mold, cooling is often added as an afterthought and is insufficient. As with the solid process, proper cooling is the key to running fast cycles. However, with the solid process, sink marks are the typical result of a hot spot. Using the microcellular molding process, the part will have blowouts.

Blowouts are phenomena where areas of the part show a swelling after the mold opens. This can occur immediately after the mold opens or over some time period after the part has been ejected. Blowouts occur when the pressure inside the cells is greater than the rigidity of the polymer matrix. When the mold opens, the confining force of the mold cavity is removed and the cells continue to expand. Blowouts occur at locations on the part that are hot on ejection from the mold. The most common areas are thick sections, sections surrounded by core pins or free standing steel, and areas formed by slides.

Good general practices for mold cooling using the conventional solid process should also be followed with the MuCell process. However, a thorough review of mold cooling should be undertaken to achieve maximum cycle time savings.

All mold inserts should have direct water lines as heat transfer across the interface of the insert and the mold base is poor. When water lines cannot be used, thermal pins should be put into the inserts and tied back to the cooling lines.

All free standing steel with height to width ratios of more than 3:1 should be cooled. Again, the preferred method is through direct water, a water line, a bubbler, or a baffle. A thermal pin may also be used but should be tied directly back to the water lines.

Water lines should also be run in all slides. These are particularly susceptible to running hot because they usually contain free standing steel.

In addition to having sufficient cooling lines, it is important to take maximum benefit of the cooling in the mold. This means that water lines should not be looped and adequate flow must be available to maintain turbulent flow.

9.8.1 Use of Higher Heat Transfer Materials

Higher heat transfer materials can also be used improve part cooling and optimize cycle times. Sprue bushings, runner plates, core pins, cooling pipes and entire core inserts have been made in high heat-transfer materials to allow the lowest possible cycle times with microcellular molding. Typically the cost of replacing a sprue bushing or core pin is relatively inexpensive when compared to the savings in cycle time. When building new molds, it is possible to consider using these types of materials for runner plates in three plate molds as well as slides and core inserts.

In each case, the molder needs to consider the temperature range to be used in the mold, thermal expansion of the alloy, as well as the hardness. While the generally lower surface hardness needs to be considered, several molders have successfully used special coatings to improve wear resistance.

The following example demonstrates the benefits of using higher heat transfer materials. Trexel molded a small blower housing with 20% talc filled polypropylene using the MuCell process. The mold had two cavities and had a valve-gated hot runner system. In standard molding, this part ran with a cycle time of 30 seconds with the cycle time being controlled by the flatness of the part. Applying the MuCell process directly to the mold, as built for standard molding, resulted in a cycle time reduction of only 3 seconds.

Cycle time reduction was limited by three mounting bosses. The dimension of each boss was 0.95 cm outer diameter, 0.32 cm inner diameter and 1.72 cm high. As cycle time was reduced below 27 seconds, these three areas would post blow on the inner surface because the core pins were much warmer than the bulk of the mold. The core pins forming these three mounting points were replaced with beryllium copper core pins prior to the next trial. After replacing these pins, the cycle time was reduced to 21 seconds. For only $300 of mold work, the customer achieved an additional 20% reduction in cycle time using microcellular molding.

Table 9.2 shows metal grades that have been used successfully in microcellular molding to optimize cycle time by improving thermal conductivity. P20 steel and H13 tool steel characteristics are provided for comparison.

9.9 Mold Finish

The MuCell molding process will not provide a high-gloss or Class A finish. Therefore, there is no benefit to building a highly polished mold. This will in fact only aggravate the tendency of the process to produce a swirled or splayed surface.

A mold with a texture or even the tooling marks will actually produce a surface in which the surface imperfections are less noticeable. This is caused by the fact that the surface roughness allows better contact with the polymer.

All features such as ribs and bosses should be draw-polished to eliminate undercuts.

Table 9.2 Comparison of Higher Heat Transfer Materials

Tradename	Description	Supplier	Hardness HRc	Thermal conductivity BTU/ft/hr/ft2/°F
Moldmax® HH	BeCu alloy	Brush Wellman	36–42	60
Moldmax LH	BeCu alloy	Brush Wellman	26–32	75
Moldmax XL	BeCu tin alloy	Brush Wellman	30	35–40
Ampco® 940	Copper alloy	Ampco Copper	20	125
QE7® Al	Aluminum alloy	Alcoa	19	91
QC7 Al	Aluminum alloy	Alcoa	17	91
7075 Al	Aluminum alloy	Several	14	75
P20 Steel	Prehardened steel	Several	28–32	20
H13 Steel	Hot Work steel	Several	52–54	15

9.10 Mold Materials

There are no additional corrosive materials or gases liberated or generated by the microcellular molding process that would require special mold construction. In fact, the reduced clamp tonnage does allow for the possibility of using softer metals such as aluminum. Aluminum works well with MuCell molding because of its significantly higher rate of thermal conductivity compared to traditional tool steels, which translates into faster cycle times.

9.11 Part Wall Thickness

Due to the nature of the microcellular molding process, there are few wall thickness limitations. As with most foaming processes, it is possible to produce sink-free parts with thick wall sections. In fact, parts have been molded using the MuCell process with wall thicknesses approaching 30 mm.

However, a unique feature of microcellular molding is its ability to produce foamed parts at a much thinner nominal wall thickness than is possible with traditional foaming techniques. To date, parts have been molded as thin as 0.3 mm while still achieving weight reductions of 4% to 8%. Since the foaming process expands to eliminate sink marks, there is more freedom to vary wall thickness for thick-to-thin sections.

As discussed previously, the maximum weight reduction possible is inversely related to the flow length over wall thickness. Thus, if higher weight reductions are desired, the flow length must be minimized or the wall thickness needs to be increased.

Trexel is focusing the MuCell process on molded parts that are 4.0 mm thick or thinner. As wall thickness becomes larger than 4.0 mm, the insulating effects of the foam start to control the cooling process. This can create a situation where the cycle time for a part molded with the MuCell process will be longer than for a part molded with the conventional solid process.

9.12 Part Ribs and Gussets

Rib designs with the microcellular molding process are not limited by an inability to pack out sink marks. In fact, the MuCell process will eliminate sink marks that normally occur in the solid molding process. This allows the rib to be designed to provide the required performance as opposed to defining processing limitations. The one limitation on rib design is that the height should be limited to no more than 3 times the nominal wall thickness. As the rib height increases beyond this, the chance of molding difficulties increases, particularly gas traps.

A couple of key points are that the draft angle should be 1 degree per side and the ribs should be draw-polished to eliminate undercuts. A radius of 1 mm at the base will improve the part performance.

As with rib design, bosses can be designed for performance as opposed to being designed to avoid sink marks. Therefore, there are a couple of key design aspects: the core pin should always extend down to the nominal wall to prevent thick sections and the core pin should be of a high thermal conductivity material. These factors will help prevent blowouts.

Venting should also be added to the boss to eliminate gas traps and also to improve the strength of weld lines on the boss.

A 1 mm radius should be added at the base of any boss. Thin bosses, i.e., those with a wall thickness of less than 2.5 mm, can be stabilized through the use of gussets. These gussets should be 80% of the nominal wall to prevent gas traps due to back flow.

9.13 Summary

Microcellular molding can use molds designed for standard injection molding. However, to maximize the benefits from microcellular molding, there are a few critical mold and part design criteria. The most critical ones are:

- Reduce the polymer mass in the sprues and runners;
- Optimize mold filling patterns through balanced runner systems and uniform flow paths in the cavity;
- Increase venting throughout the mold but particularly at the parting line and for blind pockets;
- Have good, uniform cooling paying particular attention to free standing steel and slides;
- Gate into thin part sections rather than thick part sections; and
- Minimize the material flow length in the part.

References

1. Beaumont, J., *Plastics Technology* (April 2001), pp. 64–69
2. *Plastics News* (October 15, 2001), p. 33

10 Microcellular Molding: Economic Considerations

10.1 Introduction

As discussed in Chapters 4 through 9, microcellular molding, and more specifically MuCell® molding, provides many benefits to an injection molding processor. This process and equipment technology results in an average cost reduction of 10% to 25% compared to conventional molding net of all MuCell related costs.

The microcellular foam process changes the fundamental cost structure of injection molded automotive components through four main sources of economic benefit:

1. Reduced operating costs through
   - cycle time reductions of up to 50%
   - reduced scrap and reject rates
   - lower energy consumption
2. Lower capital costs through
   - purchase of smaller and fewer machines and
   - fewer and less expensive molds
3. Reduced material costs through
   - component density reduction,
   - thinner design and
   - material substitution
4. Thermoplastic parts that are
   - flatter
   - straighter
   - free of sink marks
   - dimensionally improved

10.2 Specific Areas for Potential Savings Using Microcellular Process

The case studies in Chapter 8 demonstrated new and unique ways in which MuCell microcellular molding has reduced overall manufacturing costs. By creating a single-phase solution between the polymer and supercritical fluid, injection pressures and part weight can be reduced, clamp tonnages can be cut by 30 to 70%, and cycle time can be reduced by 30%. Obviously, these improvements can be translated into cost reductions by reducing material, improving quality and reducing cycle time.

However, other benefits are not as obvious. A few examples of these benefits are described in the following.

10.2.1 Use of Smaller Molding Machines

Purchasing smaller machines can reduce initial capital investment, especially if a tiebar-less machine is purchased. For example, in commercial applications it is possible to run molds normally used in 500-ton presses using the conventional solid process in a 200-ton tiebarless press using microcellular molding. Clamp tonnage was reduced to such an extent that the only limitations were whether the mold would actually fit between the machine base rails and whether the shot size was adequate. Additionally, a typical 500-ton machine consumes 21 to 25 kW/h, whereas a 200-ton machine consumes 10 to 12 kW/h. This is a 50% reduction in energy costs. Table 10.1 summarizes the differences outlined above in capital cost and energy expenditure for a 500-ton mold run in a 200-ton tiebarless machine.

Table 10.1 *Potential Capital and Energy Savings from Implementation of Microcellular Molding*

Process	Machine size	Capital cost	Energy usage*
Standard	500-ton	$228,000	21 – 25 kw/hr
MuCell Molding	200-ton tiebarless	$170,000	10 – 12 kw/hr
	Percent Reduction	25%	50%

* Dependent on actual application

10.2.2 Reduce Energy Consumption

Injection molding can be an energy-intensive process. With the rapidly escalating cost of electricity, customers have been looking at the energy savings potential of microcellular molding using the same molding machines as solid molding. A study performed in 1999 by W.H. Fuller & Company of Glastonbury, CT, USA concluded that the MuCell process would generate substantial savings. Using a 400-ton injection molding machine as an example, Fuller calculated savings of 298,698 kWh per year. Depending on location, the savings translate into US$ 10,000 to US$ 30,000 per year for only one machine.

Trexel also ran an actual test in March 2001. Using an Arburg 420C 42-ton molding machine, Trexel manufactured nylon tensile bars as both a solid and a microcellular part to compare the relative energy use. To record the data, a Hobo data logger was attached to the Arburg machine and programmed to record the amperage every 5 seconds. Each test lasted 30 minutes. Using the solid process, 63 tensile bars were produced.

Then the system was optimized for the MuCell process using nitrogen. Clamp tonnage was reduced from 40 to 3 tons, pack and hold time from 3 to 0.1 seconds, cycle time from 28.9 to 19 s, part weight from 45.2 to 31.2 g, and the MPP was increased from 30 to 70 bar. During this 30-minute run, 97 MuCell tensile bars were produced.

Analyzing the data provided in Table 10.2 showed that the MuCell process produced a 2% reduction in running amps. However since 97 MuCell tensile bars were produced versus 63 solid tensile bars in the same amount of time, a 36% reduction in energy use per tensile bar was measured. The MuCell parts also had significantly less warpage and far more consistent dimensions than the solid parts.

10.2.3 Reduce Colorant Usage

As with any foam process, microcellular molding results in parts appearing more opaque or whiter than parts molded using solid molding. For light color parts, the microcellular process can reduce colorant usage over the solid molding process. On the other hand, this does mean that bright or dark color parts become more difficult to make using microcellular molding.

Table 10.2 Energy Readings and Savings for Molding Nylon Tensile Bar

Solid			**31% Weight reduced (MuCell process)**		
Sample Readings			*Sample Readings*		
Date	Time	Amps	Date	Time	Amps
3/13/01	06:40:15.0	44.24	3/13/01	05:53:25.0	49.71
3/13/01	06:40:20.0	47.56	3/13/01	05:54:15.0	46.00
3/13/01	06:40:25.0	49.12	3/13/01	05:54:20.0	49.71
3/13/01	06:40:30.0	48.34	3/13/01	05:54:25.0	43.46
3/13/01	06:40:35.0	48.14	3/13/01	05:54:30.0	42.68
3/13/01	06:40:40.0	43.46	3/13/01	05:54:35.0	46.39
3/13/01	06:40:45.0	46.00	3/13/01	05:54:40.0	49.71
3/13/01	06:40:50.0	45.02	3/13/01	05:54:45.0	39.16
3/13/01	06:40:55.0	46.39	3/13/01	05:54:50.0	49.71
3/13/01	06:41:00.0	49.51	3/13/01	05:54:55.0	45.41
Average Amps		47.19	Average Amps		46.43
Average kW		37.60	Average kW		36.99
No. of Pieces		63	No. of Pieces		97
kWh / Piece		0.60	kWh / Piece		0.38
			Energy Savings / Piece		**36%**

10.2.4 Reduce Tooling Costs

Mold tooling costs can be reduced using microcellular molding. Because of the lower injection pressure and clamp tonnage requirements for many applications, molds can often be made out of aluminum instead of steel. Tooling costs can also be reduced through increased lifetime due to lower mechanical stresses imparted with each molding cycle.

10.2.5 Increase Parts per Shot

Because of the reduced injection pressure and clamp tonnage requirements of the microcellular molding process, the number of cavities in a mold can be increased. This would allow more parts to be made per shot and increase productivity. The limiting factor for implementation of this option is usually the size of the mold that can fit on the existing molding machine.

10.2.6 Reduce Blowing Agent Costs

A molder using a chemical blowing agent (CBA) to make molded foam parts can reduce blowing agent costs by switching to the MuCell microcellular process. Nitrogen costs approximately US$ 0.11/kg and carbon dioxide is approximately US$ 0.37/kg versus US$ 4.40 to US$ 12.00/kg for CBAs. Assuming 0.5% nitrogen and 0.2% CBA make similar parts, the cost of the blowing agent per kilogram plastic in a microcellular molded part is at most 21% of that for a part molded with CBAs.

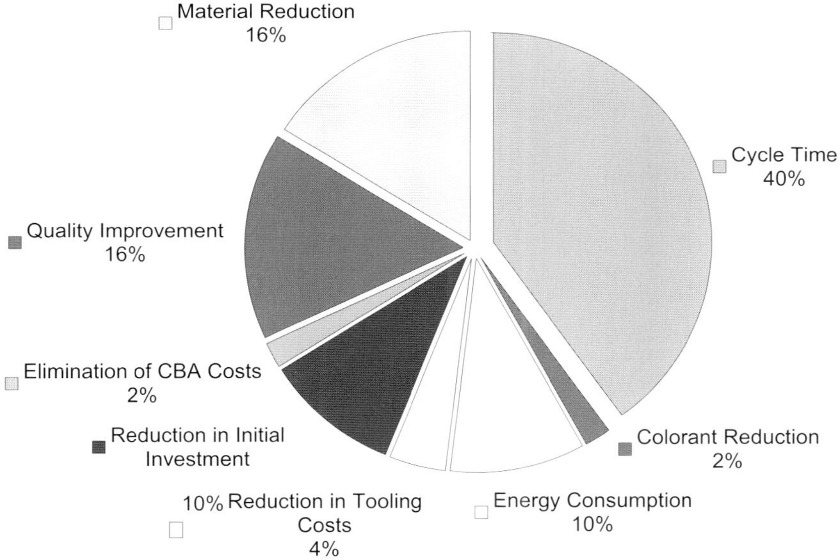

Figure 10.1 *Typical cost savings breakdown*

10.3 Total Cost Reduction Due to Microcellular Molding

Actual applications using the MuCell molding process have demonstrated typical cost reductions (see Fig. 10.1). Cycle time reduction, quality improvement, and material reduction provide the greatest benefits for the applications run to date, but other significant cost reductions include colorant reduction, energy consumption reduction and tooling costs.

As expected, most MuCell microcellular molding processors have performed a Return on Investment (ROI) analysis. Figure 10.2 shows calculated savings per year and payback time for the initial investment for some microcellular molded parts. For these calculations, the savings per year are based only on cycle time or material savings and the initial investment was assumed to be a new microcellular molding machine. Ancillary savings such as reduced energy consumption or reduced capital cost is not considered and would only increase the ROI or decrease the payback time. Upgrading an existing molding machine to the microcellular process rather than purchasing a totally new machine would also decrease the initial investment.

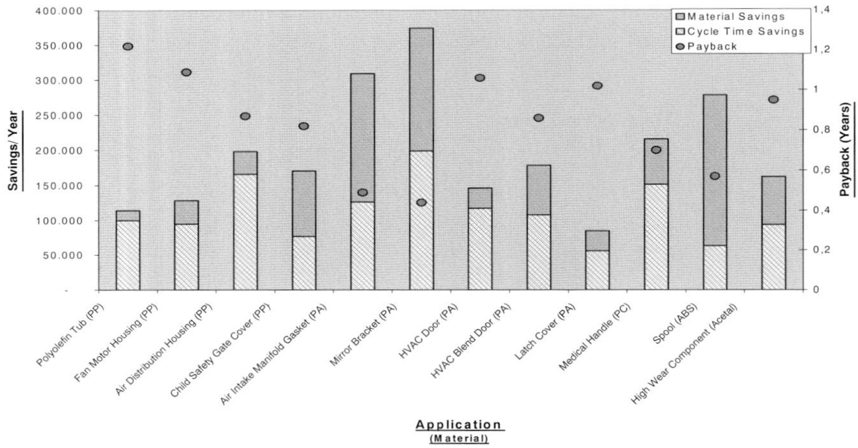

Figure 10.2 Savings and payback time in microcellular molding applications

10.4 Summary

The MuCell® microcellular molding technology is a proprietary manufacturing process of Trexel, Inc. for producing microcellular foamed plastics. The MuCell microcellular foam process uses supercritical fluids (SCFs) of atmospheric gases to create evenly distributed and uniformly sized microscopic cells throughout a polymer. Suitable for injection molding (including structural foam processing), blow molding, and extrusion, this breakthrough foam process enhances product design, improves processing efficiency, and reduces product costs. The MuCell process does not require chemical blowing agents (CBAs), hydrocarbon-based physical blowing agents, nucleating agents, or reactive components.

MuCell molding technology is available as an upgrade to installed injection molding equipment and as an option on selected new injection molding machines. All MuCell-capable machines – new and upgraded – are also capable of conventional, non-foaming solid operation.

The MuCell molding technology permits molders to reduce raw material consumption while producing strong, lightweight products, extending considerably the applications of foamed polymers. The benefits of this microcellular foam process to molders include significant weight reduction, the ability to foam thin-walled parts, reduced processing temperature, and up to a 50% reduction in injection pressure and clamp tonnage.

In addition, the MuCell process enables molders to foam materials that cannot be foamed successfully with conventional foaming technologies, such as high-temperature polysulfones, polyetherimides (PEIs), and thermoplastic elastomers (TPEs) and vulcanizates (TPVs) such as Shell's Kraton® rubbers and Advanced Elastomer Systems' Santoprene® TPVs.

11 Microcellular Molding: MuCell Process Alternatives

11.1 Introduction

As microcellular molding has become more of a commercially viable process, alternatives to Trexel's MuCell process for making thermoplastic microcellular molded products have been developed. The three alternative methods presently being touted as commercially viable for making microcellular molded articles are:

- Foam Injection Molding (FIM) by the Institut für Kunststoffverarbeitung (IKV);
- Ergocell by Demag Ergotech; and
- The use of chemical blowing agents (CBAs).

At this time, it is not believed any microcellular molded articles are produced commercially using the alternative processes to MuCell molding.

The following is a summary of each alternative based on public information. Determination of the merits and demerits of each will be left to the reader as well as determination of patent rights and coverage for all the thermoplastic microcellular molding processes.

11.2 Foam Injection Molding

Drs. Walter Michaeli and Sasan Habibi-Naini and their colleagues at the Institut für Kunststoffverarbeitung (IKV) have developed a foam molding process known as Foam Injection Molding (FIM). In papers recently presented, FIM has been shown to mold microcellular articles [1, 2].

FIM utilizes a conventional molding machine equipped with a specialized gas dosing station such as the one made by Maximator (Zorge, Germany) and a specially designed 'fluid injection nozzle' added between the plasticator and the shut-off nozzle. As shown in Fig. 11.1, the extension includes gas injectors that feed the blowing agent from the gas dosing station through internal and/or external sintered metal sleeves into the polymer melt. The molten polymer/blowing agent mixture is then injected into the mold through a specialized nozzle. This nozzle design has been patented in Germany [3]. To achieve more and better mixing of the molten polymer/blowing agent solution, IKV notes that a static mixer can be mounted between the fluid injection nozzle and the shut-off nozzle.

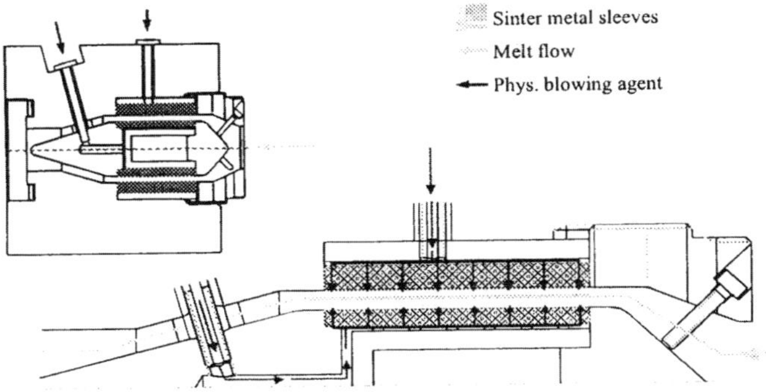

Figure 11.1 *Foam injection nozzle barrel extension [1]*

FIM has been demonstrated to work with carbon dioxide as a blowing agent in several polymers including filled and unfilled polypropylene, polystyrene, acrylonitrile-butadiene-styrene (ABS), thermoplastic elastomers (TPEs), and PC/ABS.

11.3 Ergocell

Demag Ergotech announced the Ergocell microcellular molding technology at the K-Show 2001 in Düsseldorf, Germany. The initial development design used a vertical plunger and dynamic mixer. However, the commercial design is now a horizontal plunger that is added as an extension to the plasticating screw with a blowing agent injection unit and mixer added between the plasticator and the shut-off nozzle. The basic design can be seen in Fig. 11.2, which was originally printed in *Modern Plastics* [4].

The details of this design and its operation are sketchy. However, it is believed that

- The screw reciprocates and injects plastic into a mold like a standard molding machine;
- The blowing agent is added continually into the molten plastic through the gas nozzles and sintered metal sleeve; and
- The blowing agent is intimately mixed into the molten polymer using a dynamic mixer attached to the end of the plunger and the plasticating screw.

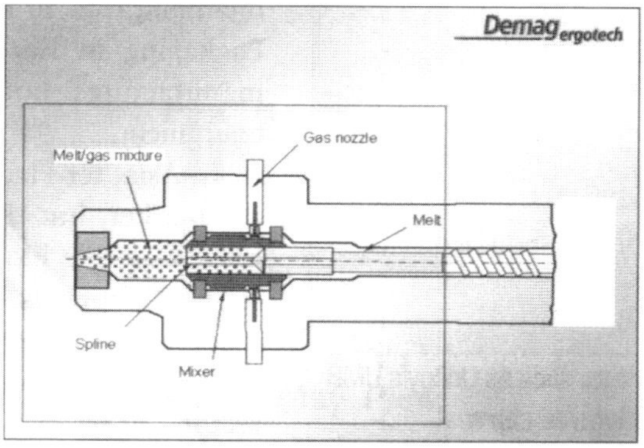

Figure 11.2 *Ergocell design schematic [4]*

Demag Ergotech has an agreement with Trexel for Ergocell users to pay a reduced license fee to Trexel.

11.4 Chemical Blowing Agents

Using the basic principles for microcellular processing developed at the Massachusetts Institute of Technology, chemical blowing agents and their concentrates have been modified to make thermoplastic microcellular molded articles. Reedy International and other CBA manufacturers have started producing CBAs and their concentrates in uniform fine particle sizes for better dispersion and more homogeneous decomposition of the CBA itself [5, 6]. Additionally, the CBAs have been formulated to decompose into carbon dioxide or nitrogen at lower temperatures to generate a molten polymer/blowing agent single-phase solution prior to injection into a mold, much like the MuCell process. Alternatively, the processing temperatures with older formulation CBAs can be increased to achieve a similar result.

Chemical blowing agents are also used as initiators or nucleators for physical blowing agents. The addition of CBAs generally reduces the cell sizes and improves the cell size distribution compared to what would otherwise be achieved without CBAs.

References

1. Michaeli, W. and Habibi-Naini, S., Paper presented at ANTEC 2002, San Francisco, CA, May 2002
2. Michaeli, W. and Habibi-Naini, S., Paper presented at Blowing Agents and Foaming Processes 2001, Frankfurt, Germany, March 2001
3. Michaeli, W., et al., German Patent DE 19 853 021 A1, issued July 27, 2000
4. Mapleston, P., *Modern Plastics*, p. 31 (November 2002)
5. Reedy, M.E. and Dudek, S., Paper presented at Blowing Agents and Foaming Processes 2001, Frankfurt, Germany, March 2001
6. Reedy, M.E., Paper presented at Structural Plastics 2002, Dearborn, MI, April 2002

12 Microcellular Extrusion

Figure 12.1 Microcellular PVC interior cladding

12.1 Introduction

Foams have been made by extrusion processing for over forty years. Conventional foam extrusion processes typically use chemical blowing agents (CBAs), hydrocarbons, or fluorinated organics such as HCFCs and HFCs. Depending on the process and the product, the blowing agent is usually a mixture of the above as well as carbon dioxide.

Standard extrusion blowing agents often add to plant and operating costs.

- Hydrocarbons like butanes and pentanes can be highly flammable. The flammability of the organic blowing agents generally leads to aging of scrap prior to regrind and to using grounded grinders; handling equipment for the blowing agents should also be spark-free and explosion-proof.
- Hydrocarbons are often regulated regarding air emissions. Foam extrusion operations using organics usually must install expensive air remediation equipment in the plants and aging warehouses. Alternatively, foam processors will often partially switch to more expensive HCFCs or HFCs to avoid plant emission requirements.
- Many of the HCFCs are addressed by the Montreal Protocol and will be banned worldwide by the year 2020 or earlier. CFCs have already been banned and many foam processors were already forced to make a blowing agent switch once.

- CBAs are expensive compared to the hydrocarbon blowing agents and it is usually difficult to control the decomposition of the CBA to generate the blowing agent. Additionally, decomposition of the CBA usually leaves chemical byproducts as well as residual CBA in the final part. Thus, in-plant reuse or recycling of CBA blown foam can be difficult.

The MuCell microcellular foam extrusion process reduces or eliminates the issues with blowing agents through the use of solely carbon dioxide or nitrogen as the blowing agent. Microcellular foam extrusion also resolves other issues that can be associated with foam extrusion:

- Large cell sizes or a wide cell size distribution make thin sheet foam extrusion difficult.
- Large cell sizes or a wide cell size distribution cause a dramatic decrease in the physical properties of the extruded foam.
- Large cells often lead to poor surface characteristics for printing or lamination.
- Conventional foam extrusion is often not able to be used in certain material types and for other materials it is limited to be used with higher-cost special resin grades.
- Many conventional foams are aged 3 to 14 days following extrusion to allow the blowing agents to escape for safe handling of the extruded foam in secondary operations.

Foam extrusion was the first continuous plastics process investigated by the Massachusetts Institute of Technology (MIT) and then by Trexel. The application of the fundamentals for microcellular batch processing to extrusion took several years of development within both MIT and Trexel. However, in time development was often performed in collaboration with processors, equipment manufacturers, and material suppliers.

Following is a general description of the MuCell extrusion process along with more information for polystyrene (PS), polyolefin, polyvinyl chloride (PVC), and thermoplastic vulcanizate (TPV) microcellular processing.

12.2 Equipment

For extrusion microcellular processing, an extrusion system must undergo some modifications as indicated in Fig. 12.2. The modifications include:

- Addition of an SCF delivery system and blowing agent source;
- Insertion of 1 to 4 SCF injectors into the extruder barrel through special taps;
- A specially-designed screw;

- Tapping the barrel and/or die head for pressure transducers and melt thermocouples as needed; and
- Die geometry modifications.

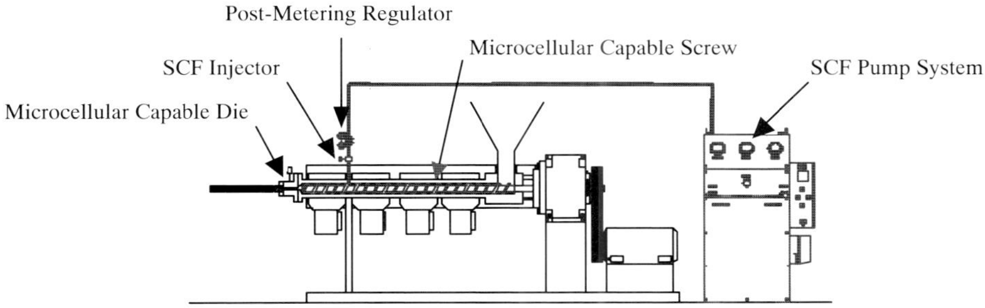

Figure 12.2 *Basic MuCell extrusion process setup*

For extrusion, the SCF delivery system includes only the SCF pump system (or SCF satellite system) with a post-metering regulator (or back pressure regulator) and the SCF injectors. A discussion of the SCF delivery system equipment can be found in Chapter 3.

The setup of the SCF delivery system can affect the consistency of the SCF flow into the extruder. The following are some basic rules to follow when installing the equipment.

- Place the post-metering regulator as close to the SCF injectors as possible.
- Keep the volume in the high-pressure SCF line to a minimum. This can be accomplished by placing the SCF pump system close to the extruder or by using the smallest tubing diameter possible for the desired mass flow rate.
- If more than one injector is used, the length of piping from the post-metering regulator to each injector should be the same.
- Install a shutoff valve after the post-metering regulator.
- Install shutoff valves before each injector if possible.

The extruder screw for microcellular processing must be designed to break the injected SCF blowing agent into small packets and mix the SCF into the molten polymer to form a single-phase solution. The screw design must also maintain sufficient barrel pressure to maintain the single-phase solution.

From feed throat to die, the basic sections of the screw are:

- Feed, transition, and metering sections for melting and conveying the plastic powder or pellets as in conventional extrusion screws;
- A blister ring or pressure restriction element to reduce upstream flow of blowing agent and maintain pressure downstream;

- A wiping section under the SCF injector(s) for breaking up and dispersing the incoming SCF blowing agent stream; and
- A mixing section for generating and maintaining the single-phase solution of molten polymer and SCF. Additionally, the mixing section may also need to cool the molten plastic when extruding certain thermoplastics.

The microcellular foam extrusion screw design is similar to the one described for microcellular molding but without the tip shutoff valve. The basic elements of the microcellular screw design are described in Chapter 4.

For microcellular extrusion, it is standard practice to place a pressure transducer just downstream of the SCF injectors; a lower cost alternative is to install a pressure gauge immediately after the post-metering regulator. The pressure transducer or gauge reads the SCF inlet melt pressure (SCF IMP) of the molten polymer in the barrel. The post-metering regulator of the SCF delivery system should be set at least 35 bar higher than the SCF IMP; maintaining this pressure differential helps to keep a consistent SCF flow into the extruder.

If it is not possible to measure melt pressure in the die, another melt pressure transducer is usually located at the end of the extruder or in the die adapter.

The melt temperature of the single-phase solution exiting the die is extremely important in microcellular extrusion. Therefore, a melt thermocouple is usually located at the end of the extruder with another optional melt thermocouple located in the extrusion die. If the exiting melt temperature is too high, the gaseous blowing agent diffuses out of the polymer too quickly, which may result in high initial foam expansion, but because the polymer cannot cool fast enough to solidify, the extruded foam soon collapses and provides little to no expansion.

Often the die melt temperature of a molten polymer is 10 to 50 °C lower with microcellular extrusion than with conventional solid polymer extrusion. However, if the exiting melt temperature is too low, then the molten polymer may have too much melt strength so the gas pressure in the nucleated foam cells cannot overcome the polymer pressure to grow, resulting in a high-density foam. If the extruded polymer is a semi-crystalline resin then a low exit melt temperature may also lead to local crystallization in the die, which may eventually lead to blockage in the die. Less dramatic, but always a concern with low melt temperatures, is the phenomenon of melt fracture, which causes poor surface finish or, in the worst cases, gross irregularities in the product.

For microcellular extrusion, the die adapter and die head must be evaluated and may need to be modified to reduce or eliminate local pressure drops. Pressure drops are undesirable as they lead to premature escape of the blowing agent from the polymer melt/SCF single-phase solution. Die melt flow channels are usually redesigned to build pressure quickly near the die exit; this allows the quickest pressure drop for the exiting polymer melt and thus induces greater cell nucleation. Die cooling is also usually evaluated so that the polymer melt can be kept at the proper exit temperature.

Die design and process parameters are intimately tied to the specific polymer being processed and to the product being extruded. As with conventional extrusion processing, each polymer/product combination requires a targeted development effort to reach commercialization.

12.3 Polystyrene Microcellular Extrusion

Extrusion of polystyrene foam was one of the first commercial development efforts for microcellular processing. Development started in 1993 and the first commercial MuCell polystyrene foam product was introduced early in the year 2000. The product was a foam-centered board or art board by Alcan Composites of Richmond, IN, USA.

The basic processing parameters of microcellular PS foam are shown in Table 12.1. Two important elements not shown here, yet critical to obtaining acceptable microcellular sheet, are sizing of the die and the design of the polymer flow path. These design elements are proprietary information of Trexel and are available only to licensees of Trexel's MuCell process.

Table 12.1 *Basic Process Parameters for Microcellular PS Foam*

Process Parameter	Setting
Extrusion system	Tandem single-screw
PS resin	Standard PS foam extrusion grades
Nucleating agent	None
Blowing agent	Carbon dioxide
Blowing agent concentration	6 – 8%
# of injectors (150 mm extruder)	4 multi-holed
Melt pressure at SCF injectors	275 – 380 bar
Die type	Annular
Die melt temperature	120 – 160 °C
Pressure at die adapter	200 – 240 bar

Microcellular PS foam can be made to similar sheet specifications as conventional PS foam. Density can be controlled from 0.030 to 0.55 g/cm^3 at extruded widths from 25 to 200 cm and potentially wider. The extruded thickness can range from 1 to 6 mm.

However, microcellular PS foam behaves differently than conventional PS foam. Table 12.2 shows some of the differences.

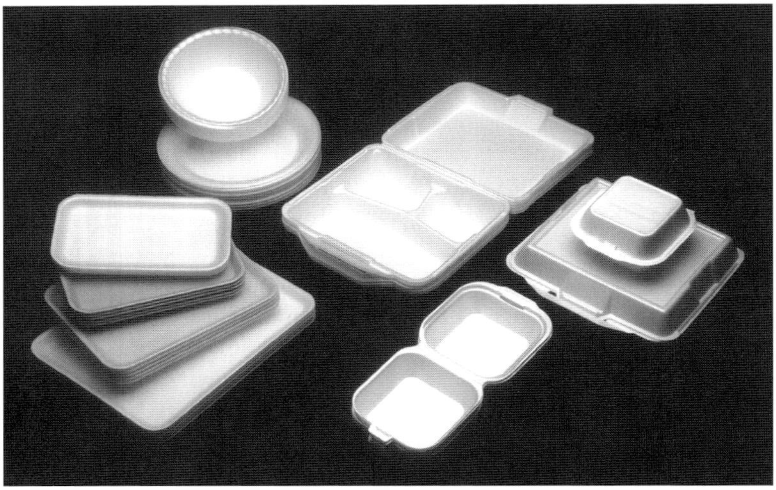

Figure 12.3 Microcellular polystyrene foam products

Table 12.2 Property Comparison Between Microcellular and Conventional PS Foam

Property	Microcellular PS foam	Conventional PS foam
Cell size, center	25 – 50 microns	250 – 500 microns
Cell size, surface	25 – 50 microns	100 – 250 microns
Ductility	Bends	Breaks
Stiffness	Flexible	Stiff
Orientation (length:width)	1:1 – 5:1	1:1 – 2:1

Figure 12.4 illustrates another difference: the surface of PS foam generated using the MuCell microcellular process is much smoother than the one generated using the conventional foaming process. The smoother surface of the microcellular product improves the sharpness of printed graphics and improves the smoothness of laminated products.

Post-extrusion operations are also much easier with MuCell microcellular PS foam. Microcellular PS foam can be thermoformed into high-quality parts immediately after extrusion; no aging is required to allow flammable blowing agents to diffuse out from the polymer and the foam cells. However, the conditions required to thermoform the same quality part will change somewhat as the microcellular foam ages for the first few days.

Conventional PS Foam Microcellular PS Foam

Figure 12.4 *PS foam surface comparison*

Because aging is not required for flammable blowing agents to exit the foam matrix, microcellular foam can be laminated soon after extrusion. Gas bubbles forming under the laminate caused by escaping blowing agent are therefore reduced or eliminated with the microcellular process.

Another advantage of the MuCell microcellular process is that scrap can be reground immediately after extrusion.

If a new plant is being built, the MuCell microcellular process also reduces the new plant costs by eliminating the need for air emission handling equipment, reducing warehouse requirements, and eliminating special equipment, procedures, and training associated with handling flammable liquids and solids.

12.4 Polyvinyl Chloride Microcellular Extrusion

As shown in the Fig. 12.1 and in Fig. 12.5, PVC can be extruded using the microcellular foam process. The interior cladding shown was produced and commercialized in Western and Eastern European markets by Dumaplast of Belgium in early 2000. Dumaplast continues to produce this product and to expand into new sales markets.

For Dumaplast, the microcellular process increased line productivity up to 25% while maintaining an acceptable surface gloss, appearance, and physical properties. Other PVC foam techniques do not work well for this application because of the 0.5 mm thickness of the ribs and walls.

A second commercial microcellular PVC product is high-end vertical window blinds introduced in 2000 by Eclipse Blind Systems. The embossed vertical slats were developed to replace existing fabric covered slats. The microcellular foam extrudate provided the proper softness for a deeper, more aesthetical in-line embossing of the product. A full pastel color palette and a wide variety of embossed designs were also provided to obtain a full product line. Other benefits for the microcellular process in making PCV vertical blinds include:

- 20-30% weight reduction while maintaining properties;
- Reduced warpage;
- Reduced contact noise;
- Meets opacity requirements;
- Stable to sunlight exposure;
- Printability; and
- Considerable freight savings as trucks were loaded to full volume instead of to full weight.

Figure 12.5 *Microcellular PVC foam products*

The basic processing parameters of microcellular PVC foam are shown in Table 12.3. Microcellular PVC foam extrusion has been adapted commercially to numerous types of standard PVC extrusion equipment including single-screw extruders and both parallel and conical counter-rotating twin-screw extruders.

MuCell PVC foam extrusion is a good alternative to the Celuka PVC foaming process and to the use of chemical blowing agents when extruding thin-wall profiles and when cell size control is essential.

Table 12.3 Basic Process Parameters for Microcellular PVC Foam

Process Parameter	Setting
Extrusion system	Single-screw, counter-rotating or conical twin screw
PVC resin	Standard rigid PVC extrusion grades
Nucleating agent	None or existing filler levels in resin
Blowing agent	0.02 – 0.25% nitrogen (generally preferred) or 0.05 – 0.50% carbon dioxide
# of injectors (70 mm extruder)	1
Melt pressure at SCF injectors	60 – 140 bar
Die type	Profile or sheet
Die melt temperature	100 – 140 °C
Pressure at die adapter	120 – 160 bar

12.5 Polyolefin Microcellular Extrusion

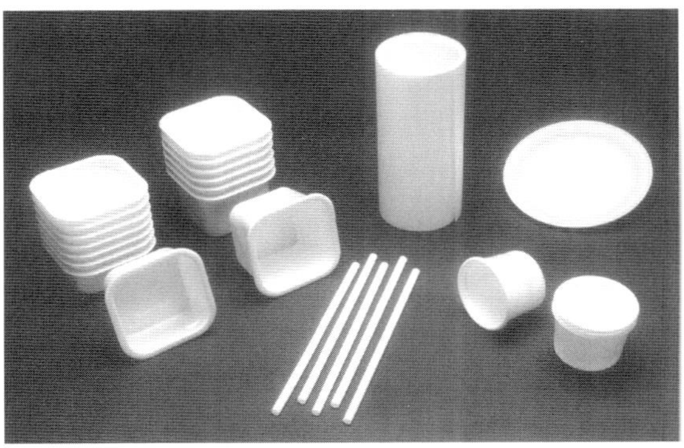

Figure 12.6 Microcellular Polyolefin Foam Products

Polyolefins such as polypropylene (PP) and high-density polyethylene (HDPE) are extruded into microcellular products without dramatic changes in process conditions. The first commercial product using the MuCell microcellular process was actually a polypropylene extruded product. A major packaging manufacturer produced and sold microcellular polypropylene straws. Up to 22% weight savings per straw was possible while maintaining line speeds. The reduction in part weight was possible without loss of stiffness by increasing wall thickness. The loss in stiffness is linear with weight reduction but the increase in stiffness is geometric with increase in wall thickness. Microcellular process technology worked in this application versus other foaming processes because of the small and uniform cell sizes produced with this technology. The major disadvantage for straws was that the material cost savings was not sufficient to justify retrofitting extrusion lines for the MuCell process. PP and HDPE can also be extruded into larger microcellular profiles with wall thicknesses up to 4.0 mm.

As shown in Fig. 12.6, PP can also be extruded into microcellular sheet and then thermoformed. As with thermoforming of solid PP sheet, the process conditions for microcellular PP sheet can be narrow; if the temperature is too hot, foam densification may result. Thermoforming equipment designed for processing PP definitely helps. Such equipment may include longer ovens, high temperature ceramic heaters, and two-stage formers. In addition to the weight savings, foamed microcellular polypropylene products can provide better insulation, for example in microwaveable containers.

The basic processing parameters of microcellular polyolefin foam are shown in Table 12.4. Microcellular HDPE foam extrusion is described in more detail in the next chapter on microcellular blow molding.

Table 12.4 *Basic Process Parameters for Microcellular Polyolefin Foam*

Process Parameter	Setting
Extrusion system	24:1 – 34:1 L:D single-screw
Polyolefin resins	Standard PP extrusion grades Standard HDPE extrusion grades
Nucleating agent	0 – 15% talc or calcium carbonate
Blowing agent	0.05 – 1.5% nitrogen or 0.15 – 4.5% carbon dioxide
# of injectors (70 mm extruder)	1 – 2
Melt pressure at SCF injectors	70 – 240 bar
Die type	Sheet, tubular, or profile
Die melt temperature	150 – 205 °C
Pressure at die adapter	55 – 120 bar

12.6 Thermoplastic Vulcanizate Microcellular Extrusion

Processing of TPVs to make foam seals, weatherstripping, gaskets, and other profiles (as shown in Fig. 12.7) is a recent effort for MuCell microcellular processing. The intention is to make replacement products for existing foamed EPDM profiles. The basic process parameters are given in Table 12.5.

Figure 12.7 *Microcellular TPV foam product*

Table 12.5 *Basic Process Parameters for Microcellular TPV Foam*

Process Parameter	Setting
Extrusion system	24:1 – 34:1 L:D single-screw
TPV resins	Standard TPV extrusion grades and TPV foam grades
Nucleating agent	None added
Blowing agent	Nitrogen (preferred) or carbon dioxide
# of injectors (70 mm extruder)	1 – 2
Melt pressure at SCF injectors	70 – 120 bar
Die type	Sheet, tubular, rod, or profile
Die melt temperature	150 – 205 °C
Pressure at die adapter	40 – 80 bar

The presently known range of basic product properties for microcellular TPV foam is shown in Table 12.6. It also shows the requirement of automotive original equipment manufacturers for weatherstripping. Compression set and water absorption were determined following ASTM D1056. The water absorption is determined after immersion of the total sample under 50 mm of water for 3 min at 17 kPA followed by 3 min at atmospheric pressure.

Table 12.6 Microcellular TPV Foam Properties

Property	OEM requirement	MuCell® foamed TPV
Density, g/cm^3	0.35 – 0.65	0.40 – 0.60
Compression set, %	< 30 – 50	35 – 50
Water absorption, %	< 5	0.1 – 5
Cell structure	Uniform	Uniform
Surface	Smooth	Smooth

The small uniform cell structure of microcellular TPV foam is shown in Fig. 12.8. For comparison, a conventional EPDM foam structure is also shown.

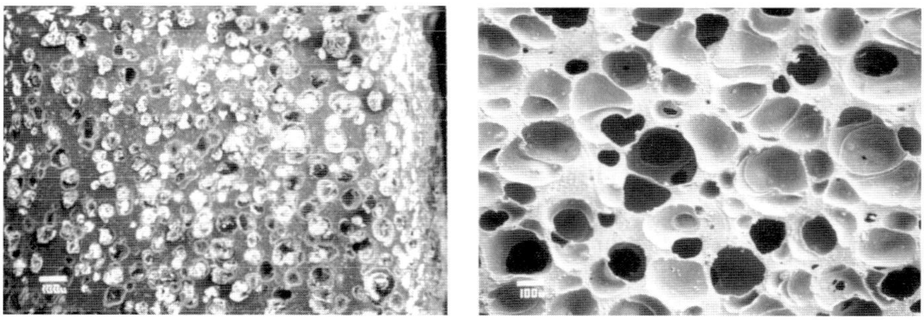

Figure 12.8 Left: Microcellular TPV; Right: EPDM

12.7 Summary

Polyolefins, TPVs, PS, and PVC are just four of the many resin types that have and can be used in the MuCell microcellular extrusion process. Other materials that have been processed include thermoplastic polyurethane (TPU), nylon, and polycaprolactone. Each material and product requires a dedicated development effort to achieve a commercially viable process and product.

The microcellular extrusion process has been used to make a number of commercial products to date, including PVC interior cladding, PVC vertical blinds, and straws. Due to the time, money, and effort required to commercialize a product using a new technology, the introduction of more extruded microcellular products may continue to be slow for the near future.

13 Microcellular Blow Molding

Figure 13.1 Microcellular blow molded bottles

13.1 Introduction

The use of conventional foaming techniques in the blow molding process has many problematic aspects.

- Insufficient control of foam uniformity results in surface defects;
- Large cell sizes result in poor mechanical properties, e.g., brittleness;
- Density reduction in the final article is difficult to achieve using normal blowing pressures;
- The cost of specialized polymers and chemical blowing agents can more than offset the savings associated with material reductions; and
- Specialized materials reduce the recyclability of the resulting articles.

The use of microcellular technology in blow molding provides a means to overcome all of these difficulties. Microcellular foams have substantially smaller cells than conventional plastic foams, are more uniform, and have higher cell densities. Microcellular foams have average cell sizes of 50 to 100 microns.

Microcellular blow molding was the third of the three major plastics conversion processes developed by Trexel. Initial material, equipment, and process experimentation and development were performed solely by Trexel. However, later development and commercial work was undertaken with a major consumer products company that led to three microcellular extrusion blow molding machines being built. The second machine was designed and built by Uniloy Milacron based on Trexel's input and design criteria; this machine was installed at Trexel as a development tool.

Microcellular blow molded articles are made using blowing agents of supercritical fluids (SCF) of atmospheric gases like carbon dioxide and nitrogen and standard grades of polymer resins. They exhibit excellent mechanical properties, solid appearing surfaces and are easily recycled. The small cells are resistant to the forces exerted during the blowing process and retain density reduction during blow molding.

Microcellular foaming in the blow molding process reduces the costs of blow molded articles by the reduction of material usage and without the increases in cost associated with expensive resins or additives. Moreover, blow molded MuCell foams are easily recyclable and thus reduce resource usage.

Like all foam processes, the major disadvantage of microcellular blow molding is that clear or transparent articles cannot be obtained.

13.2 Equipment

In extrusion blow molding, the extruder must be modified as discussed in the last chapter on microcellular extrusion processing. In summary, the extruder must be drilled to accept one or more SCF injectors, melt thermocouples, and pressure transducers. It may also be necessary to extend the barrel length, though an L:D as short as 24:1 will work in many applications.

The SCF delivery system for extrusion blow molding is the same as the one required for microcellular extrusion processing. Even though Trexel worked only on extrusion blow molding, injection blow molding with MuCell technology is also expected to be possible. The major difference would be the use of an injection molding SCF delivery system instead of an extrusion SCF delivery system. Chapter 3 details the differences between the two types of SCF delivery systems.

For standard blow molding equipment, the transfer tube and die head must be evaluated and modified to reduce or eliminate local pressure drops. Pressure drops are undesirable as they lead to premature escape of the blowing agent from the polymer melt/SCF single-phase solution. The die head used for internal trials at Trexel was a standard blow molding head employing an offset spider-design distribution system. The flow path through the head was modified to follow the design rules required to make microcellular foam structures. These designs create a high nucleation density during extrusion, resulting in small cell structures in the foam. Die annulus diameters ranging from 1.59 to 3.49 cm were adapted to this head. Ovalized tooling has also been successfully used, though the degree of ovalization required is reduced substantially in comparison to solid blow molding.

13.3 Materials

Materials used for microcellular blow molding include high-density polyethylene (HDPE) grades typically recommended for blow molding or wire coating. These materials had melt flow indices (MFI) ranging from 0.16 to 0.7 g/10 min under the standard HDPE testing conditions and specific gravities from 0.94 to 0.96. Generally, the lower melt flow index materials give better cell definition and uniformity and also contribute to a lessening of the stretch that occurs in the parison-formation stage of blow molding.

Although, HDPE grades with high molecular weight that are specified by high load MFI can usually give good microcellular, the much higher system pressures generated effectively limit their application to large part blow molding. It was speculated that this may be due to the high melt strength of the resin, which resists the formation of foam cells. Most likely the cells nucleate but collapse to higher degree than grow.

Standard fillers like talc, calcium carbonate, and titanium dioxide may also be used in microcellular blow molding. The use of fillers can drastically reduce the amount of SCF blowing agent required to form microcellular blow molded articles. The amount of filler can be optimized for mechanical properties and blowing agent level depending on the HDPE grade used. A typical filler level ranges from 5 to 15%.

The SCF blowing agent can be either carbon dioxide or nitrogen. Both provide good performance in the formation of microcellular blow molded articles. Typical blowing agent levels are shown in Table 13.1.

Table 13.1 *Typical SCF Blowing Agent Levels*

Blowing agent (BA)	Filler	BA level, %
Nitrogen	No	0.50 – 1.50
	Yes	0.01 – 0.10
Carbon dioxide	No	2.00 – 5.00
	Yes	0.05 – 0.50

13.4 Processing Parameters

The process settings of the blow molding machine extruder are largely unchanged from solid processing when using fillers. When no fillers are used, the barrel zones after the SCF injectors may be reduced by 10 to 25 °C because of the reduced resin viscosity due to the dissolved SCF blowing agent.

Die and lip temperature settings are usually ramped down towards the lip. Lower temperature settings lead to a higher pressure build behind the die and thus a better microcellular parison and article. However, the temperature settings must be maintained high enough to prevent resin from freezing in the die or from premature foaming due to lowered solubility of blowing agent in the molten polymer.

Standard parison blowing conditions may be used and optimized to obtain microcellular articles. However, programming the blow air pressure can lead to drastic improvements in part quality and properties. An initial short burst of higher-pressure air is followed by low-pressure air and then the parison is held and cooled in the mold under high-pressure air. The exact blow air pressure profile is dependent on the design of the blow molded article.

13.5 Summary

The microcellular foaming process has been used to make blow molded parisons with density reductions of more than 25% and with the characteristic small cells of the MuCell process. These parisons have uniform appearance and are free from defects such as pinholes or large cells. The thickness of the parisons can be controlled through the process conditions and the size of the exit gap in the microcellular formation die and the diameter of the parison can be varied by the diameter of the annular die. Parison diameters from 2.54 cm up to 4.4 cm have been extruded with thicknesses ranging from 0.025 to 0.24 cm.

Microcellular foam parisons made of HDPE can be extruded with average cell sizes ranging from 100 microns down to 15 microns, and with density reductions of more than 25%. Parisons blow molded into bottles can retain their density reduction and appearance characteristics.

The MuCell blow molding process can be used to make blown bottles with significant density reductions and in a variety of sizes. However, the maximum density reduction possible is usually not desired when maintaining physical properties is an objective.

In blow molded articles, the physical properties lost by density reduction may be regained through increasing the wall thickness. Usually the loss in physical properties due to microcellular processing is linear with density reduction while physical properties often increase geometrically with wall thickness. However, the typical weight reduction that can be expected in a microcellular blow molded milk bottle ranges from 5 to 10% over a solid molded bottle while retaining physical properties.

The commercial viability of microcellular blow molding must still be demonstrated. A few blow molding processors and original equipment manufacturers are continuing refinement of the process developed by Trexel.

Appendix 1 MIT Patents

U.S. Patents

Baldwin, D.F., et al., U.S. Patent RE 37 932 (2002)

Park, C.B., et al., U.S. Patent 6 051 174 (2000)

Suh, N.P., et al., U.S. Patent 6 005 013 (1999)

Park, C.B., et al., U.S. Patent 5 866053 (1999)

Colton, J.S., et al., U.S. Patent 5 160 674 (1992)

Cha, S.W., et al., U.S. Patent 5 158 986 (1992)

Martini-Vvedensky, J.E., et al., U.S. Patent 4 473 665 (1984)

International Patents

Cha, S.W., et al., European Patent 580 777 (2002)

Cha, S.W., et al., Canadian Patent 2 107 355 (1999)

Cha, S.W., et al., Korean Patent 171 911 (1998)

Cha, S.W., et al., Japanese Patent 2 625 576 (1997)

Appendix 2 About Trexel, Inc.

Trexel, Inc. is a commercialization stage venture-funded MIT spin-off with more than twenty employees. Trexel is located approximately 15 miles north of Boston, in Woburn, Massachusetts. Trexel's patented MuCell process is used to produce injection molded parts worldwide (as well as some extruded products) to provide significant cost savings and otherwise unattainable quality improvements. Due to the introduction of small, precise amounts of supercritical nitrogen or carbon dioxide gas into the molten resin during the molding process, the MuCell process enables the production of otherwise unattainable stress-free parts and parts which maintain strict dimensional stability.

Trexel's business model is based on licensing its intellectual property and know-how to molders of plastic parts and/or their end-user customers. The company has an extensive microcellular foaming patent portfolio, which provides strong barriers to direct competitors or unauthorized users. Trexel also designs, manufactures, and sells the Trexel® SCF System, MuCell® Interface Kit, and MuCell® Modular Upgrade Kit, all of which have been uniquely engineered to meet the requirements of the MuCell process.

To facilitate easy implementation of the technology on industry standard molding equipment, Trexel has established licenses with leading plastic machinery equipment suppliers throughout the world. These manufacturers offer a MuCell option for their standard injection molding machines, thereby serving as an indirect distribution vehicle while providing a platform for the MuCell technology.

To support the company's international sales activities, regional MuCell® support centers have been created by Trexel in the U.S., Germany, Japan, and Korea. In addition, there are ten worldwide locations where machinery equipment suppliers are demonstrating the MuCell process to their own customers, and Trexel's engineering group has established a network of exclusive manufacturing relationships on three continents to produce the company's proprietary precision engineering equipment.

Trexel's facility in Woburn, Massachusetts includes a 15,000-square-foot product development laboratory. The product development laboratory is complete with injection molding and extrusion apparatus, including dryers, downstream equipment, resin storage, and other auxiliary processing equipment. The laboratory is supplemented by a materials test facility for measuring the structure and performance of MuCell® products.

As of March 2003, Trexel operates a MuCell microcellular foam extrusion line in either single or tandem mode, at output rates up to 250 lb/h. Injection molding equipment, equipped with the MuCell process, is shown in table A.1.

Trexel's process and materials test laboratory has a scanning electron microscope for observing the microcellular foam morphology, a tensile/compression test for analyzing mechanical properties, as well as standard equipment including densitometers and optical microscopes.

Table A.1 *MuCell Injection Molding Equipment at Trexel*

Clamp tonnage (t)	Manufacturer	Special feature	Screw size (mm)	Max. shot size (g)
55	Arburg	Vertical and horizontal clamp	20	28
88	Arburg		30	88
200	JSW	All Electric	40	120
200	Engel	Tiebarless	60	666

U.S. Patents

Anderson, J.R., et al., U.S. Patent US 6 376 059 (2002)

Xu, J., U.S. Patent US 6 322 347 (2001)

Blizard, K.G., et al., U.S. Patent US 6 294 115 (2001)

Burnham, T.B., et al., U.S. Patent US 6 284 810 (2001)

Tupil, S.P., et al., U.S. Patent US 6 235 380 (2001)

Blizard, K.G., et al., U.S. Patent US 6 231 942 (2001)

Blizard, K.G., et al., U.S. Patent US 6 169 122 (2001)

International Patents

Burnham, T.A., et al., European Patent EP0923443B1 (2002)

Christel, A., et al., European Patent EP0991576B1 (2002)

Pierick, D.E., et al., European Patent Application EP0952908B1 (2002)

Pierick, D.E., et al., Australian Patent AU0742749B2 (2002)

Pierick, D.E., et al., Taiwanese Patent TW0424039B (2001)

Published International Patent Applications

Pierick, D.E., et al., International Patent Application WO 02 090 085 A1 (2002)

Kim, R.Y., et al., International Patent Application WO 02 081 556 A1 (2002)

Kishbaugh, L.A., et al., International Patent Application WO 02 072 927 A1 (2002)

Anderson, J.R., et al., International Patent Application WO 02 055 284 A2 (2002)

Kishbaugh, L.A., et al., International Patent Application WO02 26 485 A1 (2002)

Vadala, J. P., et al., International Patent Application WO02 26 484 A1 (2002)

Anderson, J.R., et al., International Patent Application WO 01 89 794 A1 (2001)

Blizard, K.G., International Patent Application WO 01 36 521 A2 (2001)

Chen, L., et al., International Patent Application WO 01 15 885 A1 (2001)

Okamoto, K.T., et al., International Patent Application WO 00 73 036 A2 (2000)

Xu, J., International Patent Application WO 00 59 702 A1 (2000)

Chen, L., International Patent Application WO 00 31 170 A1 (2000)

Pierick, D.E., et al., International Patent Application WO 00 26 005 A1 (2000)

Pallaver, M., et al., International Patent Application WO 99 63 019 A1 (1999)

Anderson, J.R., et al., International Patent Application WO 99 32 544 A1 (1999)

Chen. L., et al., International Patent Application WO 99 32 543 A1 (1999)

Tupil, S.P., et al., International Patent Application WO 99 04 968 A2 (1999)

Christel, A., et al., International Patent Application WO 99 00 313 A1 (1999)

Pierick, D.E., et al., International Patent Application WO 98 31 521 A2 (1998)

Burnham, T.B., et al., International Patent Application WO 98 08 667 A2 (1998)

Appendix 3 Selected Other Microcellular Patents and Patent Applications

Patents

Kanada, M., et al., U.S. Patent US 6 461 725 (2002)

DeSimone, J.M., et al., U.S. Patent 6 403 663 (2002)

Handa, P., U.S. Patent US 6 391 934 (2002)

Harris, K.M., et al., U.S. Patent US 6 386 992 (2002)

Baumgartl, H., et al., European Patent EP 707 935 B1 (2000)

Keiser, L.H., et al., U.S. Patent US 5 851 617 (1998)

Published Patent Applications

McCarthy, D.C., et al., U.S. Patent Application US 2002 0 079 320 A1 (2002)

Schürmann, E., et al., International Patent Application WO 02 247 886 A1 (2002)

Gruber, H., et al., U.S. Patent Application US 2002 0 056 935 A1 (2002)

Broome, R.C., et al., U.S. Patent Application US 2001 0 035 085 A1 (2001)

Boyer, T.D., International Patent Application WO 01 097 584 A1 (2001)

Bessette, M.D., et al., International Patent Application WO 99 047 573 A1 (1999)

Appendix 4 MIT and Trexel References

Selected MIT References

Baldwin, D.F., Ph.D. Thesis, Massachusetts Institute of Technology, Cambridge, MA, January 1994

Baldwin, D.F., Park, C.B., Suh, N.P. *Polym. Eng. Sci.* (1996) 36, p. 1446

Baldwin, D.F., Park, C.B., Suh, N.P. *J. Des. Manufac.* (1997)

Baldwin, D.F., *et al.*, *J. Japan Soc. Polym. Proc.* (1994b) 6, pp. 187-194 and 245-256

Cha, S.W., Ph.D. Thesis, Massachusetts Institute of Technology, Cambridge, MA, 1994.

Colton, J.S., Ph.D. Thesis, Massachusetts Institute of Technology, Cambridge, MA, 1985.

Colton, J.S., Suh, N.P., *Polym. Eng. Sci.*, (1987) 27, pp. 493-499.

Colton, J.S., Suh, N.P., *Polym. Eng. Sci.*, (1987) 27, pp. 500-503.

Durril, P.L., Griskey, R.G., *AIChE Journal*, (1960) 12, p. 1147

Durril, P.L., Griskey, R.G., *AIChE Journal*, (1969) 15, p. 106

Kumar, V., Ph.D. Thesis, Department of Mechanical Engineering, M.I.T., 1988

Martini, J., Ph.D. Thesis, Massachusetts Institute of Technology, Cambridge, MA, January 1981

Martini, J., Waldman, F.A., Suh, N.P., *Society of Plastics Engineers Technical Papers*, (1982) 28, pp. 674-676

Park, C.B., Ph.D. Thesis, Massachusetts Institute of Technology, Cambridge, MA, May 1993

Park, C.B., Baldwin, D.F., Suh, N.P., *Polym. Eng. Sci.*, (1995a) 35, pp. 432-440

Sanyal, J., Ph.D. Thesis, Massachusetts Institute of Technology, Cambridge, MA, June 1998.

Seeler, K.A., Kumar, V., *J. Reinf. Plast. Comp.*, (1992) 12, pp. 359-376

Shimbo, M., Baldwin, D.F., Suh, N. P., *Polym. Eng. Sci.*, 1995

Suh, N.P., *J. Eng. Ind.,* Trans. ASME, (1981b) 104, pp. 327-331

Suh, N.P., *et al.*, *J. Eng. Ind.*, Trans. ASME, (1981b) 104, pp. 332-338

Suh, N.P., in Innovation in Polymer Processing: Molding. Stevenson, J. (Ed.) (1996) Hanser Publishers, Munich

Suh, N.P., The Principles of Design (1990) Oxford University Press, New York

Suh, N.P., Axiomatic Design: Advances and Applications (2001) Oxford University Press, New York

Suh, N.P., *et al., Proceedings of the 1993 NSF Design and Manufacturing Systems Grantees Conference*, (January 1993) Charlotte, NC, pp. 315-326

Suh, N.P., Keynote speech presented at Society of Plastics Engineers ANTEC 2000, Orlando, FL, May 2000

Youn, J.R., Suh, N.P. *Polym. Comp.*, (1985) 6, pp. 175-180

Selected Trexel References

Anderson, J., *et al.*, Paper presented at Foams 1999, Parsippany, NJ, October 1999

Bernstein, D., Paper presented at Foams 2002, Houston, TX, October 1999

Blizard, K., *et al.*, Paper presented at Foams 1999, Parsippany, NJ, October 1999

Chen, L., Sheth, H., Kim, R., Paper presented at Society of Plastics Engineers ANTEC 2000, Orlando, FL, May 2000

Chen, L., Sheth, H., Wang, X., Paper presented at Society of Plastics Engineers ANTEC 2000, Orlando, FL, May 2000

Hyde, L.J., Kishbaugh, L., Paper presented at Society of Automotive Engineers Meeting, Detroit, MI, March 2002

Jacobsen, K., Pierick, D., Paper presented at Society of Plastics Engineers ANTEC 2000, Orlando, FL, May 2000

Kishbaugh, L., Paper presented at Polyolefins 2002, Houston, TX, February 2002

Pierick, D., Szczurko, D., Paper presented at Molding 1998, January 1998

Pierick, D., Szczurko, D., Paper presented at Molding 2000, March 2000

Sheth, H., Chen, L., Paper presented at Society of Plastics Engineers ANTEC 2000, Orlando, FL, May 2000

Szczurko, D., Paper presented at Molding 2002, New Orleans, LA, March 2002

Xu, J., Paper presented at Society of Plastics Engineers ANTEC 2002, San Francisco, CA, May 2002

Appendix 5 MuCell Injection Molding Process Sheet

Mold/Part: ____________ Resin: ____________

Date: ____________ Solid Shot Weight: ____________

Operator: ____________ Solid Part Weight: ____________

Process Condition #:	1	2	3	4
Time:				
SCF Delivery System/Manifold				
System Pressure:				
Pre-Metering Pressure:				
Post-Metering Pressure:				
Flow Rate:				
MFE Size:				
Part Information				
MuCell Process Shot Weight:				
MuCell Process Part Weight:				
% Shot Weight Reduction:				
SCF Injector Operation				
Injector 1				
Open Position:				
Close Position:				
Open Time:				
Injector 2				
Open Position:				
Close Position:				
Open Time:				
SCF Dosage/Cycle:				
% SCF:				
SCF Delivery Pressure:				
SCF Inlet Melt Pressure:				
MuCell Process Pressure:				
Pressure Drop on SCF Inj. Open:				
Screw Speed:				
Recovery Time:				

Machine Parameters				
Shot Size:				
Transfer Point:				
Cushion:				
Injection Time:				
Injection Pressure:				
Cooling Time:				
Cycle Time:				
Clamp Tonnage:				
Injection Profile (position/velocity)				
1	/	/	/	/
2	/	/	/	/
3	/	/	/	/
4	/	/	/	/
5	/	/	/	/
Hold Profile (time/pressure)				
1	/	/	/	/
2	/	/	/	/
3	/	/	/	/
4	/	/	/	/
5	/	/	/	/
Barrel Temperatures				
Zone 1 (Feed)				
Zone 2				
Zone 3				
Zone 4				
Zone 5				
Zone 6				
Zone 7 (Nozzle)				
Melt Temperature				
Mold Temperature				
Stationary:				
Moving:				

Appendix 6 Mucell Process Troubleshooting Chart

*** Primary variables

** Secondary variables

* Marginal variables

Possible Solution			Problem Area: Part Issues					Problem Area: Cell Structure Issues		
			Internal blisters	Surface blisters	Post-blow	Dimples	Swirls, heavy surface wrinkling, & cold flow appearance	Cell coalescence	Non-uniform cell structure	Large cells
MuCell Process Variables	SCF Level	Decrease	***		**		**			
		Increase				***			***	***
	Change SCF Type to	CO2						* >4mm thick		
		N2						* <4mm thick	**	**
	MPP	Decrease								
		Increase	***						*	
	Weight Reduction	Decrease				***			**	
		Increase								
	Check delivery pressure Δ on injector open									
	Check injector open/close settings									
	Confirm SCF is not out of solution									
Machine Variables	Mold Temperature	Decrease			***			*		
		Increase		*			*		*	
	Barrel Temperature	Decrease			**			***		
		Increase					*			
	Injection Speed	Decrease		***	*		***			
		Increase						***	***	***
		Profile		***			***			
	Injection Pressure	Decrease								
		Increase								
	Screw Recovery Speed	Decrease								
		Increase	*							
		Profile								
	Holding Time	Decrease						*		
		Increase			*					
	Holding Pressure	Decrease								
		Increase								
	Cooling Time	Decrease								
		Increase			***					
	Cushion	Bottom								
		Increase								
	Clamp Tonnage	Decrease								
		Increase								
	Shot Size	Decrease			**					**
		Increase						**	**	**

*** Primary variables

** Secondary variables

* Marginal variables

Problem Area									
Cell Structure Issues			Process Issues						Possible Solution
Cells only in thick areas	Cells only at end of flow	Large cells at end of flow	Short shots	SCF out of solution	Inconsistent part weight	Variable screw recovery time			
				***		***	Decrease	SCF Level	MuCell Process Variables
***	***	***	**				Increase	SCF Level	MuCell Process Variables
				*	* <1mm thick		CO_2	Change SCF Type to	MuCell Process Variables
							N_2	Change SCF Type to	MuCell Process Variables
						***	Decrease	MPP	MuCell Process Variables
			*	***	**	**	Increase	MPP	MuCell Process Variables
							Decrease	Weight Reduction	MuCell Process Variables
							Increase	Weight Reduction	MuCell Process Variables
				**	***	***	Check delivery pressure Δ on injector open		MuCell Process Variables
				Yes	Yes	Yes	Check injector open/close settings		MuCell Process Variables
			Yes		Yes	Yes	Confirm SCF is not out of solution		MuCell Process Variables
							Decrease	Mold Temperature	Machine Variables
**			***				Increase	Mold Temperature	Machine Variables
						**	Decrease	Barrel Temperature	Machine Variables
**			***	**		**	Increase	Barrel Temperature	Machine Variables
							Decrease	Injection Speed	Machine Variables
	***	**	***				Increase	Injection Speed	Machine Variables
*	*				**		Profile	Injection Speed	Machine Variables
							Decrease	Injection Pressure	Machine Variables
			*				Increase	Injection Pressure	Machine Variables
							Decrease	Screw Recovery Speed	Machine Variables
				*		***	Increase	Screw Recovery Speed	Machine Variables
					***		Profile	Screw Recovery Speed	Machine Variables
							Decrease	Holding Time	Machine Variables
			*		**		Increase	Holding Time	Machine Variables
							Decrease	Holding Pressure	Machine Variables
					*		Increase	Holding Pressure	Machine Variables
					**		Decrease	Cooling Time	Machine Variables
							Increase	Cooling Time	Machine Variables
					***		Bottom	Cushion	Machine Variables
							Increase	Cushion	Machine Variables
							Decrease	Clamp Tonnage	Machine Variables
		*					Increase	Clamp Tonnage	Machine Variables
***							Decrease	Shot Size	Machine Variables
		*	***		***		Increase	Shot Size	Machine Variables

Appendix 7 Glossary

ABS: acrylonitrile-butadiene-styrene polymer

amorphous material: material that does not crystallize; material softens upon heating and hardens upon cooling; often transparent; generally, viscosity change is ramped as passing through phase change; thermoplastic examples are polycarbonate and polystyrene

back pressure regulator: pressure control device that maintains constant pressure upstream

ball valve: a check valve design that uses a cylinder with a ball and seat to open and close; prevents backwards flow of material; used in some screw center check valves and in SCF injector check valves

blister ring: short screw section with increased root diameter; typically used for pressure restriction or to prevent upstream (backwards) flow of molten material

blowing agent: chemical additive used to generate foam cells in a material

blowing agent supply: tank, dewar, or blowing agent (nitrogen) generator that supplies blowing agent to SCF delivery system

blowout: surface deformation caused by part being demolded before cooling completely; has rough inner surface or elongated cell structure

bypass manifold: MuCell equipment that includes bypass valve, delivery pressure regulator, and post-metering pressure regulator; attaches to injection molding equipment; in SCF flow between pump (or satellite) system and injection valve

bypass valve: valve that opens and closes opposite to the injection valve diverting SCF flow to atmosphere

CBA: see chemical blowing agent

cell growth: the stage in foaming where nucleated cell sites grow to form foam cells

cell size: average diameter of one or all cells in a foam material

cell size distribution: distribution of cell diameters in a foam material

center check valve: valve or reverse element screw flights located prior to wiping section; reduces or eliminates gas loss in the mixing and wiping sections

chemical blowing agent: plastic additive that decomposes upon heating to generate blowing agent that is usually carbon dioxide or nitrogen

clamp tonnage: force used to keep two halves of mold closed in injection molding

Class A surface: part surface with no imperfections; surface is very smooth and glossy

cold runner: unheated channel that injected material flows through from sprue to mold cavity

cooling channel: path or conduit cut through a mold body to be passageway for coolant (usually water) to circulate and cool mold and thus molded part

critical pressure: pressure above which a gas or liquid may become SCF

critical temperature: temperature above which a gas or liquid may become SCF

delivery pressure regulator: pressure regulator that controls and maintains the pressure to the SCF injectors
dimensional stability: amount of dimensional change a molded part undergoes; usually measured over time, through thermal cycling or under load
dimples: small (1–2 mm in diameter) depressions in part
dump valve: located on SCF delivery unit; opening both valves empties high pressure lines in SCF delivery system

ejector pin: pin used to push molded part out of mold
endothermic: uses heat; as opposed to generating excess heat
EPDM: ethylene-propylene-diene monomer rubber

fault light: located on SCF delivery unit; indicates alarm condition exists on PLC
foam: material containing bubbles or holes; has reduced density compared to solid material
foam cell: bubble or hole in foam material

gas assist molding:: injection molding process in which gas is injected into the mold cavity after the molten plastic is injected; typically used to eliminate sink marks, to obtain class A surface, or reduce part weight
gas inlet valve: open/close valve at entrance to SCF delivery unit
gas pump: pump that pressurizes gas

HDPE: high density polyethylene
heterogeneous nucleation: cell nucleation occurring with the aid of a nucleation agent
HIPS: high impact polystyrene
hold phase: portion of molding cycle after packing in which plasticating screw is maintained in its forward position; time and pressure set to be high enough for injected material in sprue to cool and not flow backwards into plasticator
homogeneous nucleation: cell nucleation occurring without the aid of nucleation agents
hot runner: heated channel that injected material flows through from sprue to mold cavity
hydraulic: mechanical actuation, often using oil pressure

idle phase: portion of molding cycle after reciprocation in which the molding machine waits before injection; usually used when waiting for mold to prepare for next injection
IMP: see SCF inlet melt pressure
injection phase: portion of molding cycle in which the plastic is injected into the mold cavity; in reciprocating molding machines, the screw moves forward
injection port: barrel drilling for inserting blowing agent injector
injection sleeve: sleeve that fits over the end of the injection valve for higher SCF flow requirements with multiple small orifices; sleeve mounts directly to extruder barrel
injection valve: pneumatic or hydraulic valve that opens and closes to have SCF flow through the injector into the barrel; mounts directly to extruder barrel for low SCF flow requirements
injector: assembly of the injection valve and if necessary, injection sleeve
injector closing position: screw position or time at which injection valve closes
injector opening position: screw position or time at which injection valve opens

inlet melt pressure: melt pressure in barrel at SCF injector(s)
in-mold decorating: injection molding into mold behind fabric or film
insert molding: injection molding around item, e.g., pin or screw, inside mold; item incorporated into molded part
intellectual property: includes patents, patent applications, trademarks, know-how, and trade secrets
interface kit: interface manifold and injector(s)
interface manifold: assembly that includes post-metering pressure regulator and gauge, delivery pressure regulator and gauge, and bypass valve; mounts at injection molding machine
internal blister: surface deformation that forms over a void or a large cell; has smooth inner surface

license: legal contract allowing use of one company's intellectual property by another; usually in exchange for a royalty (monetary) payment
liquid pump: pump that pressurizes liquid
living hinge: very thin (0.1 to 0.2 mm) area between thicker walls that acts as a hinge, e.g. between container and attached lid

mass metering: metering of material flow by mass; corrects for temperature and pressure of material being measured
melt thermocouple: temperature measurement device inserted into the molten polymer stream through an extruder barrel or die body
microcellular material: foam with average cell size of 100 microns or less
mixing section: extruder screw section prior to screw tip designed for forming a single-phase solution of SCF and molten polymer
MPP: microcellular (or MuCell) process pressure; pressure maintained on the single-phase solution in front of the screw tip
microcellular process pressure: see MPP
mold vent: thin passageways cut from mold cavity to exterior of mold for passage of air and if necessary blowing agent upon plastic injection into a closed mold
MuCell: registered trademark of Trexel, Inc.

notched Izod: impact strength test using test specimen with cut notch; hammer swings and hits specimen at notch; tester measures force required to break specimen
nucleating agent: additive for initiating or enhancing cell nucleation in foam; common examples are talc and calcium carbonate
nucleation: initial stage of foaming; a few molecules of blowing agent aggregate and come out of the molten polymer to form a location where a cell will grow

OEM: see original equipment manufacturer
original equipment manufacturer: final equipment manufacturer/assembler, e.g., for injection molding machines, automobiles, and extruders
organoleptic: taste and/or odor
ovalized tooling: blow molding die with varying die gap between inner mandrel and outer die; typically outer diameter of inner die is round and inner opening of outer die is an oval with a very short distance between the foci

PA: polyamide; nylon
pack phase: portion of molding cycle directly after plastic injection into the mold; time and pressure set high enough to keep plastic in mold until sprue cools
parison: tube of plastic extruded out of blow molding machine prior to being molded into part
parting line: location where two mold halves meet on molded part
P_c: see critical pressure
PC: polycarbonate
PE: polyethylene
PEEK: polyetheretherketone
PET: polyethylene terephthalate
physical blowing agent: gas or liquid added to plastic powder or pellets at the feed throat or injected into the molten plastic in the extruder that generates foam
plasticator: extruder portion of an injection molding machine
PLC: programmable logic controller; SCF delivery unit controller
polyolefin: aliphatic hydrocarbon polymer, e.g. polypropylene and polyethylene
POM: polyoxymethylene; acetal
post-blow: see blowout
post-metering pressure regulator: pressure regulator that controls and maintains the pressure after the mass flow element; set at 3750 psi (260 bar)
pneumatic: air actuated
PP: polypropylene
PPO: polyphenylene oxide
pre-metering pressure regulator: pressure regulator that controls and maintains the pressure prior to the mass flow element
pressure transducer: measurement device that determines the pressure of the molten polymer; typically inserted through the extruder barrel or die body
PS: polystyrene
pump pressure: pressure at exit of gas pump
pump system: SCF delivery unit with pumps
PVC: polyvinyl chloride

reciprocating screw: screw that moves backwards and forwards in its barrel; in injection molding, screw rotates and melts polymer when moving backwards and injects molten polymer into mold when moving forward
reciprocation: screw moving backwards in an injection molding machine plasticator barrel
reverse screw flights: used for smaller extruder screws as a center check valve; the unbroken screw flights run in opposite direction of standard screw flights of screw
rupture disc: metal disc that breaks at known pressure

satellite system: SCF delivery unit without pumps; connects to pump system which provides SCF to unit
saturation: maximum amount of SCF that can be dissolved in the molten polymer; dependent upon polymer type, filler/reinforcement amount, temperature, pressure, and gas type
SCF: see supercritical fluid

SCF delivery system: all the equipment from the exit of the gas tank to the inlet of the extruder barrel, including SCF delivery unit, interface manifold, and injectors
SCF delivery unit: cabinet with flow control equipment and controllers; may or may not include pumps
SCF dosage control: system for determining, controlling, and maintaining consistent SCF dosage into the extruder barrel per injection molding machine cycle
SCF dosage start position: screw position at which injection valve opens
SCF dosage time: see valve open time
SCF flow control: system for determining, controlling and maintaining consistent SCF flow rate
SCF inlet melt pressure: pressure of the molten polymer in the extruder barrel at the injector
SCF outlet valve: open/close valve at exit of SCF delivery unit
screw recovery: portion of molding cycle when plasticating screw travels backwards while spinning to melt plastic and build up quantity of molten plastic in front of the screw for injection
screw stalling: injection molding screw stops moving backwards during reciprocation
semi-crystalline material: material that melts when heated and crystallizes upon cooling; often opaque; generally viscosity change is dramatic when passing through phase transition; thermoplastic examples include polypropylene, polyethylene, polybutylene terephthalate (PBT)
short shot: injection molded part does not completely fill out
shut-off nozzle: injection nozzle with a valve that is normally closed but opens during injection into the mold; allows flow only into mold
single-phase solution: homogeneous mixture of molten polymer and supercritical fluid
sink mark: large depression or concavity on the surface of a molded part
splay: streaks across surface of molded part
sprue: initial part of mold that plastic is injected into; transitions from plasticator nozzle to mold runners
structural foam molding: low pressure injection molding process for thick or large part molding; cell sizes are typically large; typically for reducing material usage
surface blister: surface deformation that forms over a void or a large cell; thin polymer layer delaminating from bulk polymer
surface defects: excessive swirls, surface wrinkling, cold flow
supercritical fluid: SCF; the state of matter above a critical temperature and critical pressure; has some properties similar to that of gas and other properties similar to that of liquid
system pressure regulator: pressure control device that maintains constant pressure downstream

T_c: see critical temperature
thermoplastic: plastic material that melts or softens on heating and hardens or crystallizes on cooling; material can go through repeated heating and cooling cycles
tip valve: valve at downstream end of injection molding screw
total shot weight: weight of all parts and runners demolded per cycle

valve gate: valve between hot runner and mold cavity; typically closed except when opens during injection

valve open time: total time the injection valve is open to allow SCF to flow into the extruder barrel

vent: opening through which SCF exits when bypass valve is open

viscosity: resistance to flow of a liquid

volume metering: metering of material flow by volume; does not correct for temperature and/or pressure of material being measured

warpage: difference in dimensions for a molded part from desired or mold dimensions

weld line: location where two flow fronts meet

wiping section: extruder screw section directly beneath injector; consists of unbroken screw flights for wiping injector face

Index